Science and Society in Latin America

In the form of a sociological pilgrimage, this book approaches some topics essential to understanding the role of science in Latin America, juxtaposing several approaches and exploring three main research lines: First, the production and use of knowledge in these countries, viewed from a historical and sociological point of view; second, the reciprocal construction of scientific and public problems, presented through significant cases such as Latin American Chagas disease; and third, the past and present asymmetries affecting the relationships between centers and peripheries in scientific research. These topics show the paradox of being "modern" and "peripheral" at the same time.

Pablo Kreimer is a principal investigator at CONICET (National Scientific and Technical Research Council of Argentina), a director of the STS Center at Maimónides University and a professor at the National University of Quilmes. He is specialized in the political and historical sociology of science.

Routledge Studies in the History of the Americas

For more information about this series, please visit: www.routledge.com/
Routledge-Studies-in-the-History-of-the-Americas/book-series/RSHAM

Science and Society in Latin America

Peripheral Modernities

Pablo Kreimer

Routledge
Taylor & Francis Group

NEW YORK AND LONDON

First published 2019
by Routledge
52 Vanderbilt Avenue, New York, NY 10017

and by Routledge
2 Park Square, Milton Park, Abingdon, Oxon, OX14 4RN

Routledge is an imprint of the Taylor & Francis Group, an informa business

First issued in paperback 2021

Library of Congress Cataloging-in-Publication Data
Names: Kreimer, Pablo, author.
Title: Science and society in Latin America : peripheral
   modernities / by Pablo Kreimer.
Description: New York : Routledge, 2019. | Series: Routledge
   studies in the history of the Americas ; 7 | Includes
   bibliographical references and index.
Identifiers: LCCN 2019009258 (print) | LCCN 2019010546
   (ebook) | ISBN 9780429266188 () | ISBN 9780367218034
Subjects: LCSH: Science—Social aspects—Latin America. |
   Latin America.
Classification: LCC Q175.52.L37 (ebook) | LCC Q175.52.L37
   K74 2019 (print) | DDC 303.48/3098—dc23
LC record available at https://lccn.loc.gov/2019009258

ISBN: 978–0–367–21803–4 (hbk)
ISBN: 978–1–03–209326–0 (pbk)
ISBN: 978–0–429–26618–8 (ebk)

Typeset in Sabon
by Apex CoVantage, LLC

Contents

Figures

Tables

Acknowledgments

First, I want to thank numerous people with whom I discussed the various issues addressed in this book over the years. In Latin America, I benefited greatly from the many conversations with Hebe Vessuri, who always encouraged me to *seguir pa'lante*, or "keep on keepin' on", and with friends from various countries, such as Antonio Arellano, Alexis de Greiff, Olga Restrepo, Rosalba Casas, León Olivé, Marcos Cueto, Léa Velho, Leonardo Vaccarezza, Daniel Villavicencio, Mariela Bianco, Yuri Jack Gómez, and Jorge Gibert, a list that abridges the endless talks and exchanges—almost always accompanied by Chilean or Argentine wine, Mexican tequila, Brazilian cachaça, or Colombian rum—I have had over the years with many other colleagues in the region.

Second, I would like to thank the co-authors of some of the texts that make up this book, particularly Juan Pablo Zabala, with whom I worked on Chagas disease for years and who has written an excellent book on its social and cognitive history. Some of the research into the origin and development of molecular biology I did in tandem with Manuel Lugones. The text on the controversies was the result of a collaboration with Lucía Romero and Paula Bilder, who felt those debates were worth studying. The research on cooperation between Latin America and Europe fell into two parts: In the first, I worked with Luciano Levin on the Latin American point of view; in the second, with Adriana Feld on the European perspective. She also worked on a very complete reconstruction of European policies toward third-country cooperation. Finally, the research on Ushuaia I did with Dalma Albarracín, one of the most brilliant people I have been lucky enough to work with. Her untimely departure was very sad indeed.

My very special thanks to Leandro Rodríguez Medina, who convinced me to publish this book and put me in touch with the editor from Taylor & Francis.

I owe an intellectual debt to various colleagues and friends from other regions around the world. To Terry Shinn and Roland Waast, I am bound by a relationship of many years' standing, ever since they were assessors on my doctoral dissertation; they have, over the years, become very dear friends. With my friends Luis Sanz and Laura Cruz, I have shared many moments—in Madrid or Buenos Aires, as well as Copenhagen or Caracas—as enjoyable as they were intellectually enriching, always accompanied by red wine. Meetings with Michel Grossetti can range from the most rigorous sociological theory to a thoroughgoing analysis of the latest football match, usually accompanied by the best cellar in Toulouse. Going to present my latest ideas in Toulouse was an often-enlightening experience. In contrast, Rigas Arvanitis and I will never stop arguing as if we were fifteen; I guess that is necessary too. Nor with "the Dominiques": Dominique Vinck, a tireless thinker and hard worker, and Dominique Pestre, who has one of the finest minds to be found where these issues are concerned. In the communication of science, I have been enriched by my exchanges with Joëlle Le Marec. Finally, spending an evening chatting with Antonio Lafuente about science or history or politics is a life experience not to be missed.

I had the adventure of organizing an STS congress with Wes Shrum in Buenos Aires, bringing together "Yankees and Latinos", and we have met constantly since to strengthen our mutual intellectual stimulus.

In my numerous extended stays in Paris, my dear friend Luc Tessier always made me feel I had a family there. The same goes for Pablo Jensen in Lyon, with whom I can argue about physics and sociology with the same passion.

I am also grateful to the researchers and PhD students of the STS Center in Buenos Aires, with whom I share my daily life.

I want to thank Ian Barnett, friend and translator, who has been editing my texts for years. In the case of this book, we had a real race against time, but we were, I think, relatively successful.

A huge thank you to my daughters, Irina and Maia, and to Adriana my wife. I needn't explain why.

1 On Peripheral Modernities, Scientific Development, Complexities, and How to Approach Them

Other Ways of Pilgrimage

Is "peripheral modernity" an oxymoron? In some ways, it could, or even, *should* be, as both terms refer to issues that seem contradictory: The peripheral is assimilated to the backward, to the traditional, to what borders on the marginal. The modern, in contrast, is typical of more developed centers, where it will eventually be able to "radiate" out to other regions. Edward Shils, who was practically the inventor of the semantic pair center-periphery, noted that

> The existence of a civilization is dependent on the existence of the centers of the societies which dominate and give character to a civilization. The relationship of central societies is constitutive of the patterns of civilization and of their growth. It is trough expansion from centers within societies that civilization are formed.
>
> (Shils 1988, p. 262)[1]

Thus, according to Shils, "members of literary, scholarly, and scientific centers regard themselves as qualified by what they think is the inherent centrality, that is, the 'seriousness' of their activities" (1988, p. 256).[2]

As a result, the dual and immediate identification of the modern with the "center", and of the backward—the traditional, the archaic—with "the peripheral" would seem inescapable. In the 1970s, these links already had been posited in terms of a certain conflict in line with the ideas of the day. One of the texts that made the biggest impact at the time was Samir Amin's (1973, p. 57) study of uneven development. For Amin, "The genesis of central capitalism constitutes the *first main manifestation* of the *law of uneven development of formations*". For Amin, this law takes the following form:

> [A] formation is never superseded from its centre but from its periphery. The main contradiction of a formation which defines the

dominant mode characterizing it is not the main aspect of a contradiction. That is to be found in another field: the periphery of the system.

(Amin 1973, p. 374)

However, when we look at the development of a globalized (modern, Euro-American) science and its deployment in less developed countries, this issue looks far more complicated. To be clear from the outset: The scientific elites of "peripheral" countries are deeply imbued with the values, practices, beliefs, and ideas of "science", although their mode of integration does not come about in the same way as legitimate members of the Euro-American community.

This is why I have borrowed the concept of the contradictory and provocative title of a book analyzing the transformation of the cultural field in Buenos Aires between the 1920s and 1930s. It acts as a condenser to help us approach these kinds of tensions.[3] Ever since I read it thirty years ago, I have been fascinated by the apparent contradiction of modern vs. peripheral. Indeed, I believe that the dynamics of science can provide a lens every bit as interesting as the study of cultural and artistic development, especially given how unexpected, or even counterintuitive, the parallel between science and culture seems. While the notion of "universal culture" has a distinctly archaic aroma, it is still extremely common—indeed, firmly anchored in common sense—to read claims to "universal science". In fact, if one tries to construct objective scientific knowledge, it can only be universal. Naturally, the issue has, for many years, been the subject of debate by the "two cultures" articulated by C. P. Snow in 1959: namely, *literary culture* and *scientific culture*.[4]

As my investigations relate to Latin America, though primarily based in Argentina, a brief description of its situation is in order: Within the countries globally considered "peripheral", Argentina was one of the first to adopt the foundations of the modern imaginary in different spheres of social life. It can be seen both in the expansion of Argentina's education system since the end of the nineteenth century, which allowed the highly effectively incorporation of waves of immigrants from Europe at the turn of the century and in the growth of institutions, the expansion of its middle classes, the development of research, and the installation of new fields of knowledge. All this was mediated by a cosmopolitan, pro-European elite in a far more pronounced way than elsewhere, both within the continent itself and in relation to other peripheral regions. Certainly, none of this erased its peripheral condition, but rather threw into contrast a considerable sociocultural development coexisting with the objective restrictions of its relative position in the world, which marked a singularity that is particularly interesting as an object of study. In strict terms of the scientific dynamic, Argentina achieved a development in certain fields in the opening decades of the twentieth century that brought a Nobel Prize winner in Medicine in 1947 (Bernardo Houssay) and another in 1970, awarded to one of his students (Luis F. Leloir).

This apparent rampant contradiction between modernity and periphery, a certain ambition embedded in a habitus of society (a particular feature of Buenos Aires, but one probably extendable to several other cities): that of "being in fashion", of following the trends deemed novel in the broad cosmopolitan space, makes the study of the emergence of new scientific fields in Argentina a particularly interesting space of observation.

It took many years, in fact, and is still certainly an unfinished task, to convince not just the general public but even—and especially!—colleagues in the social sciences of the epistemic equivalence between the study of science and other social objects. To put it another way using a quick example, the concept of "social class" or of "poverty" is as constructed as that of "gene" or "atom": They all have the same epistemic status, highly dependent on the locally situated context, on the culture that produces them, and so on.[5]

Another inspiring idea I gleaned from reading many years ago, when working on my doctoral dissertation, Michael Mulkay's book, subtitled *A Sociological Pilgrimage* (1991). In it, Mulkay makes a selection of his texts written over twenty years and markedly reflective in tone. He charts four stages in his intellectual development. Through his texts, he shares with us how both his objects of theoretical analysis and his convictions slowly morphed and changed.[6]

The idea of pilgrimage nicely sums up the search for different ways to tackle the issue of science in its different aspects, facets, demonstrations, processes, and consequences. Given that scientific knowledge is a complex, multiple object grappling with the natural world, which it seeks to explain, and with the social world that permeates it and which it also transforms, our strategies to study it and try to explain it tend also to be many and varied.

Like Mulkay, I also intend here to chart a quarter-century of my investigations into science, its social determinants, and its consequences. However, my journey is certainly different from Mulkay's: In the first place, because I was not subject to many paradigm shifts over the course of my research, having joined social studies of science (SSS) when they were already well established; and second, because my target was always oriented to Latin America and, as I discuss in Chapter 2, there is a great difference between analyzing "mainstream" science and peripheral science. Let us look briefly at the implications of these two issues.

My first concerns were oriented to studying science-related policy issues and, in particular, to understanding the role of science policies. This choice seemed natural thirty years ago, since the so-called "Latin American Thought in Science, Technology, and Society" developed in the 1960s and 1970s had put the focus precisely on political issues over and above their sociological dimensions. From the authors of the period, like Jorge Sábato, Amílcar Herrera, Máximo Halty, Francisco Sagasti, or even Rolando García, the one who immediately caught my attention was Oscar Varsavsky. Not so much because of his radical stance toward social change and the role of science in a society—in those days, everyone imagined themselves marching almost inevitably toward a socialist revolution—but because he was, by far, the one who had the greatest sociological—and epistemological—insights.[7] Varsavsky was the first—at least, the first I read—to mention how scientific agendas were imposed from the centers of greatest international power, while questioning (as early as the 1960s) the emerging system of evaluation and organization of scientific careers through the sacralization of the "paper".

Still interested in science policy issues, I had soon embarked on my doctorate in Paris, under the direction of Jean-Jacques Salomon, who had been one of the pioneers in these topics and the creator of the science and technology policy division in the Organisation for Economic Co-operation and Development (OECD). We immediately struck up a friendship, establishing a kind of filial relationship—he was the age of my father, who had died a few months earlier—which, oddly enough, I later took as an object of analysis (filiation structures in scientific research was one part of my doctoral dissertation and the title of an old article).

Jean-Jacques was an intellectual in the broadest sense (a French sense!) and had interacted with characters like Jean-Paul Sartre (I remember him telling me about his job as a young editor on *Les Temps Modernes*), Georges Canguilhem, and Raymond Aron (these last two had acted as advisors on his doctoral thesis, which culminated in his 1970 book *Science et politique*, although, in my opinion, Aron was the greater influence). More than anything, he was extremely broad in his criteria, and he had a highly varied teaching staff in his STS Center at the Conservatoire national des arts et métiers (CNAM). On this staff was Bruno Latour, then a young lecturer who, alongside Michel Callon, had been editing the CNAM's bulletin *Pandore*, one of the first publications in this emerging field, and had just moved to the Centre de Sociologie de l'Innovation (CSI) of the Ecole nationale supérieure des Mines de Paris. The French translation of *Laboratory Life* had been published very recently and was followed hot on the heels by *We Have Never Been Modern*. After reading the first of the books, which I must have finished in a night, I said to myself, "This is what I want to do: get inside laboratories and observe how scientists produce knowledge". (Years later, I found out the same happened to other young sociologists.) So, I forgot about science policy for many a year and began to see how I could work on these issues. At the same time, I wrote my thesis on another of the topics I was deeply interested in the formation of the STS field, in this case in France (funnily enough, the subject was suggested to me by Latour himself). The topic of how a look at the sciences from the social sciences was constructed I had always found very interesting; indeed, a few months ago, I coauthored an essay on the subject in Latin America (Kreimer & Vessuri 2017)

Sheer coincidence led me into molecular biology laboratories, one in the Institut Pasteur in Paris, another in Birkbeck College, London (the laboratory set up by John Bernal!), to "work *à la Latour*".[8] Having completed that part of my research, I had to decide how to continue and decided that I was not interested in studying those laboratories if I could not compare what was going on there with similar institutes in Latin America. So I made the contacts to visit a large institute in Buenos Aires and another in Rio de Janeiro, the question being to see what changed in two flagship European centers (both had been created by and had hosted a number of Nobel laureates) and two elite centers located in two peripheral countries. Although I had, in the end, to suspend my visit to the Brazilian laboratory, being in molecular biology laboratories for a prolonged period of time proved to be nothing short of a revelation. Shortly after my research, I dropped the "Latour method" and began to research other issues. I was not so interested in actants, hybrids, and rhetorical devices, as I was in other issues, like intergenerational relations (the construction of traditions through intergenerational filiation), how the use of knowledge is articulated, the relationship between the modes of organization of work and sociocognitive products, the types of laboratory management, and so on. Then, on arriving in Buenos Aires, the relations between hegemonic and peripheral centers became the subject that would occupy me, in one form or another, in years to come.

Here I come to the second difference with Mulkay's pilgrimage. Like almost all researchers in the STS field from the central countries, he devoted himself to analyzing a very particular type of local knowledge, the kind belonging to the mainstream of international science, often called Northern or Western science: radio astronomy, the economic analyses of the British health system, the debates in the British parliament on assisted fertility techniques and human embryo research, and biomedical research in oxidative phosphorylation in the United Kingdom and the United States, naturally.[9]

In this respect, Sandra Harding correctly advances the thesis that the project to break with an idea of value-free or neutral science showed, through the ethnographic studies, that "Northern sciences, too, are 'socially situated', taking their problems, ontologies and other background assumptions, and preferred methods of collecting and evaluating data from their local historical contexts" (Harding 2008, p. 140). And therefore, "this work showed how Northern sciences, the most acclaimed achievements of the Enlightenment legacy, were nevertheless in distinctive ways still 'ethnosciences'" (Harding 2008, p. 141).

However, despite Harding's warning, the fact that practically all of the most cited authors from the sociology of science in the past few years make little reference to this side of their theoretical proposals is no less of a paradox. By contrast, they seem to postulate that their descriptions apply fairly generally to a variety of contexts. For example, theories such as actor-network-theory (ANT) (Callon, Latour, Law), The Third Wave and the studies on experts and expertise (Collins, Evans), Discourse Analysis (Mulkay, Gilbert, Ashmore), and Coproductionism (Jasanoff) appear to present no limitations in terms of the social, economic, geographical, or cultural context. Starting with a "provincial" study (Law & Lin 2017), they become universal. This is curious, because that privilege seemed to be the monopoly of physics or chemistry, at least until the arrival of constructivist perspectives in the 1970s.

Unfortunately, this was how many researchers from Latin America also saw it, and they applied these categories without asking how suitable they were to study and to explain the (markedly different) processes seen in their own societies. If this is the case, it is because the same type of relationship of cognitive domination we have observed in the "hard" sciences (which is, in part, the object of this book) is also seen in the field of the social studies of science. John Law (Law & Lin 2017) recently agreed with this diagnosis: "STS is dominated conceptually, linguistically, bodily, metaphysically and institutionally by provincial EuroAmerican and especially English-language practices".

My "pilgrimage" therefore could never resemble Mulkay's and his Euro-American colleagues', because he does not need to explain its limited, local character, as he deals with hegemonic science. In contrast, to the extent that I dealt with Latin America and, therefore, with peripheral science, I had to justify that provincial, subaltern character and, of course, its relevance to the global field of the social studies of science. An example will help us better understand the question: When Pierre Bourdieu speaks of the genesis and structure of the literary field, he does not feel it necessary to explain that he will be setting out to study the *French* literary field or to limit Gustav Flaubert's role in that process:

> We know how much Flaubert contributed, along with others, notably Baudelaire, to the constitution of the literary field as a world apart, subject to its own laws. To reconstruct Flaubert's point of view, that is, the point in the social space from which his vision of the world was formulated, and that social space itself, is to have a real chance of placing ourselves at the origins of a world whose functioning has become so familiar to us that the regularities and the rules it obeys escape our grasp.
>
> (Bourdieu 1992 [1996], p. 48)

No reference to the local: The genesis of the French literary field is the genesis of the literary field *tout court*. But Bourdieu is far from being an exception. Latour and Woolgar, in what almost became a bestseller of the social studies of science, point out:

> It might also be objected that the work of the particular laboratory we have studied is unusual in that it is relatively poor at the intellectual level; that its activity comprises routinely dull work, which is not typical of the drama and conjectural daring prevalent in other areas of scientific work. However, the Nobel Prize for Medicine was awarded to one of the members of our laboratory in 1977, soon after we began preparation of this manuscript. If the work of the laboratory is merely routine, then it is possible to receive what is perhaps the most prestigious kind of acclaim from the scientific community for the kind of routine work we portray.
>
> (Latour & Woolgar 1979, p. 32)

Nowhere in the book is there any hint that their analysis of the construction of scientific facts is performed in one of the most prestigious laboratories in the world: namely, La Jolla, California. What is more, the word "California", a "situated" place of knowledge production, appears only two times in the entire book: one time, in tangential reference to the requirements of the University of California, and the second time to affirm how scientists marvel at the "universalization of knowledge": "How extraordinary that a peptidic structure discovered in California works in the smallest hospital in Saudia Arabia (sic)!" This, however, by no means calls the authors into question.

Is There Life After *Laboratory Life*?

After ending my own laboratory life in Paris, London, and Buenos Aires and having completed my doctorate in France, I felt enough was enough. I spent my first few weeks at the Institut Pasteur trying to follow the prescription of "naïveté" suggested by Latour to the letter, and even his recommendation of agnosticism and methodological ignorance, or "estrangement". It fitted me like a glove: I was totally ignorant of everything being researched there. But that presented me precisely with a problem, because I was not able to turn my ignorance into knowledge; I mistook the refrigerated cells where they kept the DNA gel for the refrigerators where they kept their sandwiches for lunch. The very expression "run DNA" brought to mind a bunch of guys in white coats chasing bits of genetic material through the streets. So I adopted the more sociologically useful idea of developing native competences (Collins 1983): I stopped wasting time and enrolled in a basic course in molecular and cell biology in the University of Paris, and then another two after that, until I was able at least to have a reasonable conversation on those topics.

So when I arrived at the institute I studied in Buenos Aires, it was my third immersion in a molecular biology laboratory, and I had been familiar with the field for some years. On completing that study and, with it, the whole comparative research work, two issues were still pinning me down: On the one hand, the contrast between the European laboratories and the Argentine laboratory (whose leaders had spent several years in Europe and maintained close ties) brought me face to face not only with the difference of contexts but also with the ways they influenced the production of knowledge and, above all, how they inserted themselves differentially in "world science". On the other hand, I began inquiring about the relatively new discipline of molecular biology, founded in Argentina in the late 1950s by a sort of "national hero", Dr. César Milstein, who had received the Nobel Prize in 1984 for the development of monoclonal antibodies. The fact that Milstein had been living permanently in Cambridge, United Kingdom, since 1966 (the year of a military coup in Argentina) added drama to the question but did not decrease the chauvinism of Argentines, who in any case trumpet being the only nation in Latin America to have three Nobel Prize winners in science (naturally, they include Milstein).[10]

So, in the following years, I devoted myself, on the one hand, to looking for ways to understand what was then called "peripheral science" (Vessuri 1983) and, on the other, to undertaking a historical reconstruction of molecular biology as an emerging field in Argentina. Here, again, I came up against "peripheral modernity": As we shall see in detail in Chapter 5, the first molecular biology laboratory in Argentina was set up in 1957, just four years after the publication of the famous article by Watson and Crick describing the structure of the double helix for DNA. In fact, there were, in those days, very few molecular biology laboratories in the world: a few in the United States, England, France, and not much more. But the one in Buenos Aires was dismantled just five years later as a result of political intervention.[11]

Aside from investigating the issues around the peripheral dimension, the study of the history of molecular biology led me to other pathways. One of these was the emergence of scientific fields as a problem of analysis and a line of empirical inquiry. In fact, even though this brings us back to the social and cognitive organization of science beyond the spaces of the laboratories and other sites of knowledge production, I observed that it was a matter quite unvisited—not to say abandoned—by the social studies of science.

The other pathway was to begin looking into Chagas disease, a tropical disease that only exists in Latin America. In effect, the parasite that causes Chagas disease (*Trypanosoma cruzi*) was one of the first objects of study around which molecular biology took shape to act as a scientific field in both Brazil and Argentina.

Studying the production of knowledge about Chagas disease led me, in turn, to question the social uses of scientific knowledge in more depth, both in general and, especially, in the peripheral regions.

So, in this game of Chinese boxes or Russian dolls, the problem of the use—and, particularly, the *nonuse*—of locally produced knowledge pointed me toward a fabric of international relations without which it was impossible to understand the processes of scientific development.

＊　＊　＊

I pointed out earlier that my meeting with Jean-Jacques Salomon was crucial. In a sense it was a continuity of my political concerns, so gearing them to the study of science policy was the most natural path. His influence went further, however. When I was seduced by—and became passionate about—the sociology of scientific knowledge, Salomon, who had trained with Canguilhem and was far more conservative in epistemic issues, said to me: "I am very open and tolerant. As long as you are rigorous and demonstrate everything you say, you can choose whatever approach you like most". At the same time, that ignorant, arrogant, young Argentine felt he was continuing the French tradition, the lineage of Aaron and Canguilhem.

My second crucial encounter was, as I mentioned, with *Laboratory Life*. While the fever lasted just a few years, they were very intense and marked me for a long time to come. By the time I arrived in Paris, Latour had already moved out of the CNAM to the Centre de Sociologie de l'Innovation (CSI), but even so I was able to attend some of his seminars discussing the thesis projects of the doctoral students from the Ecole des Mines. I went many times, but never dared to speak at the group meeting; the students of Latour, Callon, and Akrich (among whom were several friends of mine) exuded a very intimidating air of superiority.

My third encounter was just as important as the others: With Hebe Vessuri I discovered how to approach peripheral science. By the mid-to-late 1980s, I had a short contract with the regional office of UNESCO to catalog courses in Latin America on science and technology policy and management. I met Hebe when I visited the Campinas program in Brazil. She was coordinating the Science Policy Department at the time, and I had already read the occasional text by her. Since then we have kept up a relationship of constant affection and exchange. We see each other around the world, often playing the role of "enlightened Latinos": she as "the Lady" of Latin American science studies, and I as "the youngster" (a condition I was progressively abandoning with the years). Off the top of my head, in addition to nearly every Latin America country, we have run into each other in some unusual places, like Gothenburg or Nairobi, Paris or Bangkok, Toluca or Cyprus.

When it came to defending my doctoral thesis—several decades ago now—and it was proposed I invite "a Latin American", I had no hesitation and put Hebe's name forward. Over the years, she has been a juror for most of my doctoral students, and even for theses directed by my students ("my little grandchildren", as she once called them). For several years now, we have also been indulging ourselves in writing a half-dozen texts together, including two recent ones: a history of science in Latin America (a long, long article that will someday become a book) and a reflective look at social studies of science and technology in Latin America (Vessuri & Kreimer 2018; Kreimer & Vessuri 2017)

The Issues We Tackle in Peripheral Modernities and Model Organisms

So, I have selected some topics from my pilgrimage which I consider essential to understanding the development and role of science in Latin American societies. We may start by asking ourselves very directly—as, incidentally, no few governments have—"What use is science in countries where more than a third of the population lives in poverty?" and then, "Should those resources not be spent on meeting more basic needs than the amusement—generally harmless, granted—of its scientists and intellectuals"?

These questions, albeit brutal, are not unreasonable. Still, I chose a different path: to show the different tensions permeating the production and use of knowledge in these countries, prioritizing questions over answers, trying, at the same time, to produce some theory, some interpretation, even to examine hegemonic theoretical perspectives in the study of science, and so analyze the extent to which they are or are not useful in order to map those "other" contexts.

The issues, then, pick up on the argument I summarized earlier: How do new disciplinary fields emerge? What kind of relationships exist between the construction of social problems and the formulation of scientific problems? How have relationships between centers and peripheries changed in recent decades? I analyze some socioscientific disputes, as well as the eternal tensions between the production of international knowledge versus the social utility of knowledge.

* * *

All the sciences have, over time, developed various *models*, which may be constructed in very different ways. One classification, among the many possible, could establish conceptual models, in vivo models, in vitro models, or in silico models (Griffiths 2012); each fulfills different functions in the knowledge production process. I am particularly interested, however, in a type of in vivo model used by biology, namely model organisms:

A model organism is a species that has been widely studied, usually because it is easy to maintain and breed in a laboratory setting and has particular experimental advantages. Over the years, a great deal of data have accumulated about such organisms and this in itself makes them more attractive to study. Model organisms are used to obtain information about other species—including humans—that are more difficult to study directly.

(Twyman 2002)

The most widely used and extremely well known are the fruit fly (*Drosophila melanogaster*), the bacterium *Escherichia coli*, and the plant model *Arabidopsis thaliana*, as well as other old acquaintances, like the zebrafish (*Danio rerio*), and obviously the *Mus musculus*, our very own house mouse.

Similarly, some years ago, I chose my own "model organism" for sociological research: Chagas disease. Let us take a quick look at some of its characteristics: Chagas disease is caused by a parasite, *Trypanosoma cruzi,* and affects between 18 and 25 million people in Latin America, which is why it is recognized as the region's main endemic (WHO 2000).

Latin American populations had been carriers of the parasite for centuries, and some texts even suggest that Inca mummies had already been infected (Fornaciari et al. 1992).[12] But until the early twentieth century, it had remained an invisible entity that sickened and killed without a name or, therefore, a complete existence.

It is essentially a "disease of poverty", its main form of infection being through the *vinchuca* (*Triatoma infestans*), an insect that nests in the walls and ceilings of *ranchos* (houses made of adobe and straw), and the rural population in endemic areas is the most affected (Briceño León 1990).

The lack of external symptoms, employment discrimination suffered by those infected, and the poverty of the majority of patients (usually from rural areas), furthermore, make Chagas disease a "neglected disease": International laboratories carry out no research and development (R&D) into new drugs because, given the characteristics of the market and the research effort necessary, it is unprofitable for these firms.

At the same time, Chagas remains a recurrent topic on the public agenda in Argentina, Brazil, and other countries and has been the subject of different policy plans since the 1950s. These plans (health, epidemiological monitoring, scientific research, housing) have led to numerous activities that, though insufficient for its eradication, have created a certain "density" of social actors articulated around the disease; in other words, they have turned Chagas disease into a public problem.

Where scientific research is concerned, Chagas (in the various approaches to the disease, its pathogen, vector, patients, and so on) is an important topic of inquiry (particularly, in recent decades, in the field of molecular biology), and several of the groups that emerge from the prestigious biomedical tradition form part of the scientific elite of the country and cluster around it. In fact, in both Argentina and Brazil, research into Chagas disease has been considered a "successful case of scientific development in the periphery", to the extent that it receives "broad recognition for its relevance and legitimacy" from the scientific community (Coutinho 1999).[13]

In what sense can Chagas disease serve as a model? In the first place, because as an issue that exists solely in Latin America, it allows me to observe the relatively autonomous ways whereby Latin American societies have viewed it over the years, designing strategies to treat the sick, eradicate it, and so on. In fact, it is impossible to find *vinchucas* or *Trypanosoma cruzi* in nontropical climates, as these organisms do not survive in cold temperatures. Although cases of Chagas disease have today been found in the United States and in certain European countries (Spain, France), this is due to migration (generally of poor rural people) and not to any possible vector-borne transmission. In other words, it is impossible to import "turnkey" solutions as in many other issues; rather, historically, it has been about how societies have constructed it as a problem and how they have tried to resolve it.

Second, there is a crossover between science and politics: scientific knowledge has played a fundamental role among the ways in which the problem has been constructed in the public sphere, from the first doctors at the start of the twentieth century to the genomics at the start of the twenty-first. And, as I argue in Chapters 3 and 4, the way to build a public problem epistemically defines the modalities of intervention in the same operation. To put it another way, whoever defines the terms of a problem has the legitimacy to define how to approach and resolve it. In the case of Chagas disease, the problem could be construed as a health, social structure, housing, insect population control, pharmaceutical, insecticide chemical industry problem, and so on. And every approach involves the mobilization of knowledge, disciplines, different actors, and institutions. Here, I would echo John Law (2004, p. 14) that "in its practice science *produces* its realities as well as describing them"; this is to say, it is *performative*.

Third, Chagas disease helps me show how a dispute emerges and is resolved in a peripheral context. Unlike in the more advanced countries, the capacity to produce "regulatory science"—science used for regulatory purposes (Jasanoff 1995, 1998)—is extremely weak in less developed countries for at least two reasons: on the one hand, due to a certain relative weakness of locally produced science, which normally needs international certification to be legitimized; on the other hand, due to the tendency of the state decision-making machinery itself, which generally does without any locally produced knowledge. For example, in the field of health, World Health Organization (WHO) recommendations are usually taken as a "black box" and applied almost without discussion in Latin American countries. However, several authors have pointed out very emphatically that "we found that many strong recommendations issued by WHO are based on evidence for which there is only low or very low confidence in the estimates of effect (discordant recommendations)" (Alexander et al. 2016).

Finally, taking Chagas disease as a model is also useful to show the tensions between an "internationalized" knowledge (the molecular biology of *Trypanosoma cruzi*) that becomes autonomous ("purified", as we shall see) from its context of production to become common currency in the international knowledge market (symbolic goods), but, at the same time, finds justification in the way of constructing a public issue.

Sociologists specializing in methodology might object that this is not strictly an analogy with a biological model, but simply a case study. There is no doubt that this type of interpretation is possible or that Chagas disease can be thought of as a singular case through which a set of cognitive and social dynamics can be observed and from which there emerges a grid of analysis that could be applied to other cases of knowledge production in the periphery. I would like to apologize to these methodologists in advance, but I prefer to go on using the analogy of the model in that it allows me to demonstrate that my interest is not in this particular case (just as molecular biologists are not interested in mice or fruit flies), but in certain structural mechanisms that transcend this disease and these actors.

The International Dimension and Method Issues

In the newspaper this morning, I read that the dollar rate in Argentina is about to soar because Turkey has devalued its currency. I also read that international prices of soya, Argentina's main export product (which has, unfortunately, replaced our traditional and far tastier beef; we are clearly, and in a variety of senses, living in vegetarian times) increased last year due to the drought in the United States. On the other hand, Latin American countries believe they can export food to certain African countries, given Africa's possible economic growth as a consequence of heavy Chinese investments. However, to paraphrase Koyré, many in this region seem to be thinking in terms of a "closed world" and refusing to see the "infinite universe".

There is a high density of studies about the development of science and technology in Latin America, and they have addressed various aspects, such as the emergence of national science systems (after the colonial period) or the development of fields scientific (Bourdieu was very successful in this territory), disciplinary, or problem focused. Many more have set about understanding the relationships between science and industry, concerned about the role of universities and influenced by the relatively successful perspective of the triple helix (Etzkowitz & Leidesdorff) or "Mode 2" (Gibbons et al.). Bibliometric research has also been done to observe changes in patterns of publication, as well as many studies on the development of science and technology policies. Other works, especially more recently, were aimed at studying various socio-technical disputes—particularly environmental ones—or understanding the development, or rather the difficulties of development, of industrial sectors. They also aimed to identify so-called "social technologies" (a contradiction in terms in that it supposes that there are "nonsocial technologies").[14]

But an overwhelming majority of studies are focused on the study of these issues as if they were closed spaces, or as if all those issues were not informed by a far broader international dynamic, which goes far beyond national boundaries. However much interest is created by things occurring in various regions of the planet; the crucial thing is that if we do not take these dimensions into account, our intended analysis of "science and technology in Latin America" will be very limited and, ultimately, unrealistic.

Let us take a brief look at a list of questions that cannot be answered unless we turn to what I have called "the international dimension": In the analysis of S&T policies and instruments in recent decades, how much influence does the most advanced countries' strategy have in attracting scientists and engineers from developing countries? What is the scope of the research priorities of, for example, the European Union? What are the results of postdoctoral residencies among young researchers at prestigious centers in the United States and Europe? How many resources come from international cooperation? What kind of turnkey technologies are imported, and which would it be possible to develop locally? How far is the traditional knowledge used—but not paid for—by transnational corporations taken into account? What are the implications, in these days of Big or Mega Science, of scientific equipment needing to be imported without the possibility of negotiation or adaptation to local needs? What capabilities do local states have to develop knowledge and question international regulations?

So, aside from Chapters 7, 8, and 9, which deal directly with international dimensions, this entire book could be thought of as an attempt to break with the illusion of a closed space sheltered from the influences if "outsiders". Naturally, this is not a question of tastes and preferences but of analytical rigor: Science has claimed its universality for several centuries, and through the social studies of scientific knowledge, we have done much to demolish that mystified image. We should also take note today of the hyperglobalization we are governed by and offer readings that allow us to design better interpretations and, therefore, better modes of intervention. We should not ignore it, like a child covering its eyes and saying, "I'm not there" and then opening its eyes and saying, "Here I am".

* * *

Where methods are concerned, I have no intention of provoking any great methodological debate.[15] I am interested, however, in using two sources of inspiration to try to give this book meaning. On the one hand, there is John Law's idea (2006) about pinboards. According to him, the pinboard is an analytical method which attempts to engage with the "messiness" of reality by articulating its complexity, diversity, and noncoherence which are all typically erased in traditional narrative accounts.

To put it another way, whereas both the social sciences and, in particular, STS studies have accepted that social reality is complex, multidimensional, diverse, and heterogeneous, with multiple actions occurring simultaneously and tensions pulling in different, often contradictory, directions, the stories continue to be linear, with a degree of diachronic continuity, and frequently dealing with a single dimension (or only a few dimensions) at a time. Leaving aside the rich, but not entirely fertile, debates and contributions around reflexivity (let us remember the fourth principle of the Strong Program proposed by David Bloor in 1976), the sociological narrative has tended to flatten out these diverse complexities.

This has always reminded me of Jorge Luis Borges's fabulous tale about Funes the Memorious,[16] an individual with a perfect memory who remembered with photographic precision every event he observed, albeit with the loss of any ability to conceptualize: Funes could not link "the dog at three fourteen (seen from the side)" with the "the dog at three fifteen (seen from the front)". Funes could thus reconstruct exactly what had happened throughout the day, and its reconstruction, of course, took him another day.

Borges's magnificent insight has, however, one drawback: If Funes's perfect memory enables him to photographically record everything, reconstructing—relating—a day should take him much longer than that. Imagine, for example, that Funes takes an hour-long bus ride. There is the reconstruction of the journey: the streets he took, the traffic lights he stopped at, the other vehicles he passed, the fluidity or density of the traffic, and so on. There is also everything that happens during the journey: a neighbor walking the dog, children on their way to school, people out shopping, flowers on balconies, arguments in bars, and so on. And, of course, we have everything that happens on the bus: the driver scratching his nose or checking his phone, the couple kissing on the back seat, the lady arguing with her son, the passengers getting on and off the bus, reading their books, painting their nails (*everyone* constantly checking their phones, naturally).

All this occurs simultaneously, and Funes should record it, including the clothes everyone he saw was wearing, as well as their stains or imperfections. Yet he does not, and Borges therefore conceals from us that the story of Funes, though he cannot conceptualize, nevertheless involves a level of selection within that chaotic world.

In that sense, Law's pinboards are a fertile tool to approach an explanation of the chaotic nature of the social (and physical) world. Yet virtually no one has attempted to replicate these methods, which seem to hold so much promise.[17] It may be that with this approach of Law's something similar occurs to what happened with reflexivity in the early days of the sociology of scientific knowledge (SSK): Many people found it attractive to apply the same rules of social causation and so on to sociology itself. And this sparked a major wake-up call about how knowledge is formed. But this pathway they soon found bothersome and, as Collins and Yearley noted, ultimately "leading nowhere".[18]

It is possible, then, that pinboard approaches bring out substantive epistemological and methodological issues, although their adoption is bothersome and keeps us away from our main objective. Yet I believe there is an option and that that pinboard is only constructed over time with a form closer to the prism or polyhedron, which shows us different faces and various angles in relation to certain issues. Broadly speaking, this is the approach of this book, where I try to open up different problems, but with a kind of evergreen conviction that, at the end of the day, I am talking all the time about the same issues. In this sense, each chapter can be read as a contribution to the completion of the image, to strengthening the (always partial) gaze on certain aspects that overlap with previous ones.

The Structure of the Book

As I pointed out in the preceding paragraphs, each of the book's chapters aims to open up a problem relating to science and its relationship with society in Latin America. These texts were written over twenty-five years, although in a variety of forms and in different languages.

Chapter 2 addresses the general problem of studying science in "non-hegemonic" contexts and, within that problem, two issues that are to me the most relevant. The first is to do with the set of relationships informing the production and use of knowledge and, on that basis, I question the appropriateness of a level of analysis that—to paraphrase Merton in the presentation of his middle-range theories—allows us to incorporate international dimensions without thereby forgetting laboratories.[19] Here, in passing, I discuss the problem of disciplines and other forms of social and epistemic organization, a subject which seems to have received little attention in recent decades in the STS field. This is striking because, toward the end of the twentieth century, disciplines were radically transformed: The landscape of the division of labor across disciplines seems to have had virtually nothing to do with the one we saw toward the end of the Second World War, for example.

Also in this chapter, I discuss the structure of relations framing science in nonhegemonic contexts, as well as ways to approach it, from Shils's earliest reflections to recent postcolonial proposals. The chapter ends with a question mark as to whether it is still possible to speak of "scientific centers and peripheries", whether this reflects the complex world we live in, or whether it is necessary to think of new ways of capturing these realities.

Chapters 3 and 4 take Chagas disease as an analytic model in order to bring out a variety of dynamics around science and knowledge in these societies through this tropical ailment. The first of these focuses on the construction of public problems and the way in which scientific knowledge is actively involved in the construction of such problems. As I noted earlier, the ways of constructing a problem are not neutral so that every approach, every disciplinary field, every focus constructs both an epistemic object and an object of social intervention, but in addition, the mode of construction determines the range of the modes of intervention.

Chapter 4, on the other hand, presents a more theoretical discussion, showing the limitations of various approaches circulating in the international mainstream of the STS field to account for these problems. Accordingly, I analyze four different approaches in the form of divertimentos, or fictions, and conclude by proposing an alternative analysis that may, I hope, overcome some of these difficulties.

In Chapter 5, we move full on to the history of science: I reconstruct the history of molecular biology in the region from its inception in the mid-1950s to the most topical issues of genomics today. As I explained earlier, I eventually came to this discipline after studying various laboratories in France, England, and Argentina at the microsocial level. I understood then that reconstructing its entire development would allow me to analyze a series of topics, such as the actors entering and exiting, the displacements in space and time, the epistemic leaps, the various forms of social organization, the institutions, and the links to policies and politics. At the same time, the study of molecular biology is anything but trivial: over this whole period, the field "colonized" most of the life sciences. If, in the 1950s, the Nobel Prize winners Houssay and, above all, Leloir, were capable of denying the importance of this emerging field, claiming it was "a mere set of techniques", a half-century later, molecular biologists came to occupy a place of hegemony and displaced most of the more "traditional" approaches or subdisciplines.

Chapter 6 is also historical and looks again at Chagas disease, but in a different form: the eruption of a dispute over whether or not to treat chronic patients with the drugs available at the time. This analysis allowed me to show the ways in which the actors—doctors, biologists, authorities, industrial laboratories—position themselves facing therapeutic discord. Like any dispute, the arguments revolve around the construction of the evidence. But in this case, this also allows us to observe the weaknesses of regulatory science in peripheral contexts and the precariousness in public decision-making grounded in fragments of knowledge whose legitimacy is fraught with problems.

The last three chapters focus on various aspects of relations between centers and peripheries. Chapter 7 reviews the process of institutionalization of science in Latin America from the point of view of internationalization, as well as the successive moments or stages of the process, from the postcolonial period and the installation of national scientific institutions to more recent times characterized by mega-networks and a new form of international division of scientific work. I note here the shifts in the training of researchers and the way careers are structured, as well as the dilemmas raised by these ways regarding the social or economic use of locally produced knowledge.

In the next chapter, the magnifying glass enlarges far more to show, in particular, the (growing) participation of Latin American individuals and research groups in European projects over the last decade. On this point, I take a two-way perspective, questioning a group of Latin American leaders who participated in European consortia and European scientists who coordinated major research consortia. Among many other issues that interest me here, possibly the most important is the modes of recruitment of non-European researchers and, above all, the distribution of tasks within these consortia.

The last chapter is also framed by international linkages, but here I take a specific case, namely the structure of linkages deployed in Ushuaia, Tierra del Fuego. The case is particularly interesting, because, years ago, I discovered that, despite being frighteningly far away from the most important urban centers, a research center based there had the highest rate of international collaboration in Argentina. Naturally, I wanted to find out why, and this led me to observe a special type of scientific research, which I have termed "research with geographical priority". This research brings into play complex negotiations in relation to access to (material and cognitive) resources in order to enforce local skills and the strategic value of a certain kind of knowledge.

* * *

Most of these texts are new to the reader in English, either because they were originally published in Spanish, Portuguese, or French, or because they were presented at conferences or other meetings, or because they were published as chapters in books with narrower circulation. Some simply remained unpublished.

Naturally, with one exception (in Chapter 5), I excluded all topics and articles originally published in English. None of the texts has been published in its original wording. Although I set out to rewrite as little as possible, I immediately abandoned the attempt. The result is that the texts—particularly the oldest—were subjected to ethical and esthetic reworking in some depth. The chapters are as follows:

Chapter 1 is previously unpublished and was written exclusively as an introduction to this book.

Chapter 2 is one of the newest texts. It contains parts of the introduction to the book I had published in Spanish in 2016, entitled *Contra viento y marea. Emergencia y desarrollo de campos científicos en la periferia: Argentina, segunda mitad del siglo XX* [Against All Odds: The Emergence and Development of Scientific Fields in the Periphery: Argentina in the Second Half of the Twentieth Century].

A previous, much truncated version of Chapter 3, coauthored with Juan Pablo Zabala, was published in Spanish in the journal *Redes, Revista de Estudios Sociales de la Ciencia* in 2006. In this version, I specifically added the whole discussion of the purification and the universalization of biological processes.

The three divertimentos reproduced in Chapter 4 have remote origins in the chapter of a book I published in Spanish in 2011, entitled *Desarmando ficciones* [Disarming Fictions]. This new version is rather more informal and provocative than the original.

Chapter 5 is a synthesis of various earlier texts. It is a broad summary of my book *Ciencia y Periferia. Nacimiento, muerte y resurrección de la biología molecular* [Science and Periphery: The Birth, Death and Resurrection of Molecular Biology], published in 2010. The description of the early days of molecular biology is based on an article published in *Science, Technology and Society* in 2002. The most recent part of the story was inspired by a chapter I wrote with Hugo Ferpozzi, which appeared in *Contra viento y marea* in 2016.

The controversies over the treatment of patients with Chagas disease reproduced in Chapter 6 is the result of a work I wrote with Lucía Romero and Paula Bilder, published in Spanish in the journal *Asclepio* in 2010.

A preliminary draft of Chapter 7 on the processes of internationalization of Latin American science and the dilemmas surrounding the use of knowledge was published in Spanish in the book *Ensamblando Estados* [Assembling States], published in Bogotá in 2013 and edited by Olga Restrepo Forero.

A mixture of texts lies at the origins of Chapter 8: on the one hand, an unpublished text I wrote with Lucíano Levin, based on a survey of Latin American leaders; on the other, the European view appeared in an article I wrote with Adriana Feld, which was published—in Spanish—in the journal *Historia, Ciencias, Saúde—Manguinhos*.

Last, a much shorter preliminary draft of Chapter 9, written with Dalma Albarracín, was presented at the Annual Meeting of the Society for Social Studies of Science in San Diego in 2013.

Notes

1. Although the idea may seem very close to the one articulated by Basalla (1967) about the spread of Western science, Shils's analysis is much finer and, above all, more rigorous from a sociological point of view.
2. In economic terms, the issue had already been raised in Latin America by Raúl Prebisch of the Economic Commission for Latin America and the Caribbean (ECLAC) and later by various authors who shaped it toward the 1970s in the so-called "theory of dependency".
3. Namely *Una modernidad periférica. Buenos Aires 1920–1930* by Beatriz Sarlo (1988). I have to point out that, while this is a brilliant book of literary and cultural criticism, it has little to do with my own approach to the development of science in peripheral contexts, possibly because of the gulf that exists between scientific and literary production. I therefore confess to drawing inspiration more from what the title suggested to me than in its theoretical implications.
4. Well known as it is, it is worth recalling a brief passage from Snow: "Literary intellectuals at one pole—at the other scientists, and as the most representative, the physical scientists. Between the two a gulf of mutual incomprehension—sometimes (particularly among the young) hostility and dislike, but most of all lack of understanding".
5. I have recently discussed this topic to analyze the relationship between the social studies of science and social science in general, where I speak of an "unrequited love". See Kreimer, 2017.
6. As his first texts date from the early 1970s, Mulkay immersed himself in the debates over the reading of Kuhn, as well as the emergence of the Strong Program (Bloor, 1976) and the constructivist sociology of science. For a more detailed discussion, see my book *De probetas, computadoras y ratones* [Of Test Tubes, Computers, and Mice] (Kreimer, 1999a).
7. Varsavsky was a chemist and mathematician who attempted to apply mathematics to social modeling. Feld (2018) hypothesizes that those "intuitions" might, in fact, originate in discussions of that period among sociologists like Gino Germani or epistemologists like Gregorio Klimovsky.
8. At a Christmas party I once attended by chance, I found myself talking to an anthropologist who ran a European-funded project on various European laboratories viewed "from outside" by non-Europeans. It was a two-way anthropology project invented by Umberto Eco, who was the president of the institution running the project and who was, in those days, doing trials in Italy and China. Thanks to that chance encounter, I was able to spend a year in each laboratory; at first accompanied by a Chinese anthropologist and later alone.
9. As we will see in various parts of this book, both expressions are confusing: for example, Mexico or Egypt is in the Northern Hemisphere, and the whole of Latin America is in the Western Hemisphere. They seem rather to be euphemisms used to avoid speaking of underdeveloped or peripheral regions. John Law, Anderson, and others refer more directly to "Euro-America". The

expression sounds closer to the truth, but is very vague even so: It leaves out Japan and includes many European countries far from the scientific mainstream: Speaking very generally, Greek, Portuguese, or Romanian science is more similar to the science done in the advanced countries of Latin America than to German, English, or French science.

10. In January 1999, I was able to interview César Milstein for an entire day in Cambridge, United Kingdom. Considering this was my first interview with a Nobel Prize winner, he never once made me feel it was such an honor: Milstein was a simple man, who spoke slowly, using an Argentine Spanish typical of the 1960s and still drinking *yerba maté* with sugar after living in England for forty years. Up close, he looked nothing like the "hero" conjured up by those who knew Houssay, the region's first Nobel Prize winner in the sciences.

11. The complete story of the birth, death, and resurrection of molecular biology can be found in my book published in 2010. See Kreimer (2010d).

12. It may be interesting to stress a parallel with Latour's text (Latour, 2000) on the Pharaoh Ramses II and his supposed death from tuberculosis, in other words, from Koch's bacillus. Similarly, I note that four molecular biologists from the Institute of Pathological Anatomy and Histology (University of Pisa, Italy), with the help of an Argentine researcher from the "Fatala Chaben" Institute of Parasitology (specializing in Chagas) had to perform a number of procedures to identify the parasite in the mummy.

13. I discuss the notion of "success" in Chapters 3 and 4.

14. For a critical analysis of the development of the social studies of science in Latin America, see Kreimer and Vessuri (2017).

15. There have been brilliant examples of this discussion in STS studies, of which the book by Law (2004) is, at least in my opinion, possibly the most interesting.

16. Let us not forget that Borges provides us with a wealth of useful reflections for the STS field. In the introduction to my 1999 book *De probetas, computadoras y ratones* [Of Test Tubes, Computers, and Mice], I could not resist the temptation to quote a paragraph from his astonishing story "Del rigor en la ciencia" [Of Exactitude in Science] about the extent of the map of the empire. Recently, hunting down references on model species in biology, I discovered that systems biology commonly refers to the "Borges Dilemma", with just this story in mind. Moreover, Latour and Woolgar, in the second edition of *Laboratory Life*, invited readers to consult Borges's "Pierre Menard, Author of the Quixote" if they wish to compare the two editions of their book.

17. Craige (2015) is one of the few who have tried to put this into practice systematically to study telemedicine in the United States.

18. Collins and Yearley proposed a sociologically interesting or—to me, at least—satisfactory solution for the dilemma of reflexivity: "Natural scientists, working at the bench, should be naive realists—that is what will get the work done. Sociologists, historians, scientists away from the bench, and the rest of the general public should be social realists. Social realists must experience the social world in a naive way, as the day-to-day foundation of reality (as natural scientists naively experience the natural world)" (Collins & Yearley, 1992, p. 308).

19. Merton (1968), discussing the sociology of mass communication and the sociology of knowledge, opposed European sociology, which dealt with the "big problems" but with little precision, and American sociology, which dealt with minor, far less relevant issues "with a great deal of rigor".

2 Studying Scientific Development from Latin America

Problems of Definition, Levels of Analysis, and Concepts

Continuing with questions related to social studies of science in a peripheral context already mentioned in the previous chapter, here we will focus on two aspects which I consider to be crucial. First, discussing the most appropriate level of analysis for our study of the production and use of knowledge in these spaces makes sense and enables us to make some general considerations which think beyond our specific case. Indeed, our investigation should situate itself between the micro level of the laboratory characteristic of research carried out in the 1980s and the macro level of a country or whole region frequent in bibliometric approaches. Second, we establish what types of concepts and modes of analysis—utilized in the context of countries belonging to the "center"—are useful and which are not to give an account of science in the peripheral contexts, as well as explaining, at the same time, what kind of theoretical production is necessary.

Therefore, the first section of this chapter is dedicated to discussing a question to which the social studies of science across the world have paid less and less attention: the organization of sociocognitive collectives. Whether we are discussing disciplines, fields, research areas, regimes, or other concepts, in my consideration the question is far from being resolved. This could be because the available concepts are not necessarily satisfactory, but more importantly it involves "shifting sands", in a process of constant change and mutation. For example, from the second half of the twentieth century on, a set of research fields and disciplines has emerged and challenged the existing structure which seemed to be a solid as a rock: Ecology, nuclear physics, the marine sciences, informatics, molecular biology, and biotechnologies are some examples. Furthermore, around the end of the last century and the beginning of the current one we have witnessed the rise of new configurations such as the neurosciences or nanosciences which are challenging the scientific structure that had seemed very robust (Bozeman, Laredo, & Mangematin 2007; Meyer 2007).

In the second part of the chapter I will discuss a thorny issue: the asymmetries in marking the production of scientific knowledge between the more advanced "centers" and the less developed "peripheries". This question, unlike the mentioned disciplinary organizations, was neglected for decades and only recently have some Euro-American academics started to investigate this question, as we shall see later.

Between Disciplines, Fields, and Arenas: The Question of the Level of Analysis and Analytical Perspective or the Relationship Between Social and Intellectual Factors

From a historical point of view, it is very interesting to observe the role that the history of a field or scientific discipline has on the dynamic of that field. That is to say, the effects it has on the constructions of lineages, traditions, and feelings of belonging.

On the uses of the history of disciplines, the analysis of Lepenies and Weingart (1983) is of particular interest. They point out that there are in fact different groups which produce histories of science, each aimed at different audiences and each fulfilling different functions. First, they identify the function of legitimation, which has two quite distinct aspects, both produced by the practitioners of a given discipline: a focus on the "wider public", aimed at portraying the "heroic" character of the pioneers or founders of a particular discipline. In some cases, these histories aimed at a mass audience are framed within the struggle for the legitimation of a new field seeking to make a place for itself among the already established disciplines. These studies can be likened to activities of scientific popularization, taking the form of disseminating the significant findings of a particular discipline. The other aspect also plays a role of legitimation, but within the discipline itself, aimed at the researchers belonging to the discipline, as well as students and new recruits, in order to retrospectively demarcate its own hegemonic traditions within its confines.

If the previous dimensions are considered to be "internal to" the field, there are other "external" histories, particularly those produced by the social studies of science (SSS):

> [B]y undermining the aura of heroic achievements and the sanctity of elitism they transmit an image of science as an everyday, social activity which is not aloof from challenges of democratic accountability. Thus, the "social studies of science" are not really "histories" of disciplines but systematic analyses which focus on the conditions of the historicity of scientific development.
>
> (Lepenies & Weingart 1983, p. 14)

Both aspects are intersected by that of professional historians of science—often educated at a basic level in the field they study, or receive an educational grounding in the relevant field after their initial academic career in order to better understand the technical, theoretical, or epistemological questions—who distance themselves from internal histories and attempt to connect disciplinary histories with broader political, economic, or cultural history. This aspect seems to be "in between" perspectives "from within" and the viewpoints or analyses "outside of" the discipline in question.

Within this penetrating analysis there is a paradox which its authors seem not to have noticed: Despite the diverse functions that the disciplinary historiographies fulfill, beyond the complex analytical distinction between the "professional historiography of science" and the "social studies of science", both currents—particularly SSS—consider the histories which emerge from the field itself to be just one more source of material for analysis. Therefore, they do not attribute to it *a priori* greater truth value than the claims of the practitioners of any other social field. In this case, it is a question of using the reconstructions coming from a particular discipline as sources which enable us to portray the actors at a particular historical moment.

The concept of "discipline", thanks to its deep-rooted implantation in university education, is understood by everybody, both "hard" scientists (physicists, biologists, astronomers, and others) and "soft" scientists (social sciences and the humanities), by the institutions of science policy and even by people outside of the scientific sphere. That is to say, it is understood by the "hard" and the "soft" as well as the "al dente".

However, and in spite of its supposedly neutral or "natural" character, the concept of discipline, like any other way of conceiving of some form of organization—in this case we are faced with the additional problem that it is simultaneously social and cognitive—corresponds with a specific determined conception of science and its concomitant practices, a particular mode of segmenting the physical, natural, or social world.

Of course, the most difficult aspect to show the constructed character of a firmly established discipline is the existence of a group of practitioners who regard themselves as belonging to this general "umbrella" and who identify, often with very strong bonds, with this definition. This is what anthropologists often refer to "native definitions", meaning that a social group identifies itself as "physicists", "chemists", or "radio astronomers". Like ships inside bottles, it would seem that the disciplines "have always been there",[1] and we should not question the reasons behind this particular partition of the world into scholars of different disciplines.

We could say, with Heilbron, that the

> standard image of a disciplinary order is that of a universe neatly divided in a large number of slightly overlapping areas, each of them being the specialty of a particular group of professional experts. The territorial metaphor captures an essential feature of the disciplinary regime: numerous similarly organized and relatively autonomous

units coexist, each with the function of producing and reproducing a specific body of knowledge about a certain part of reality.

(Heilbron 2004, pp. 23–24)

This metaphor fits well with most native definitions, which feel comfortable with this simultaneously epistemic and social division of the world.

To express this in a slightly different way, belonging to a discipline seems to guarantee its members a corpus of intellectual questions or problems, considered a legitimate object of study, within the general distribution of problems they are concerned. At the same time, from a static and idealized view, this division correlates to a social organization articulated by a collective identification, and, therefore, a disciplinary group is that which shares a set of problems to investigate—a region of the broad world of investigative possibilities—along with a body of relatively coherent theories which lay the groundwork for understanding and a social organization which functions as an institutional support guaranteeing would we could call the "simple reproduction" of the discipline. This takes place by means of a process of educating new generations, in university departments, through obtaining resources from funding research agencies (generally also organized along disciplinary lines), in professional associations, conferences, and so on. Ultimately, this model is firmly anchored in the well-known Kuhnian definitions of paradigm, which in one of its many meanings refers to a set of relatively articulated theories which govern normal science, and in another definition refers to "that which a scientific community believes in". We should bear in mind, however, that by the time of writing the postscript Kuhn had already partly abandoned the concept of paradigm (object of numerous critiques and debates) and suggested instead that of disciplinary matrix (Kuhn 1970).

Rudolf Stichweh, a Luhman disciple, provides a good synthesis when he proposes that the modern scientific disciplines are an "invention" of the end of the eighteenth and the beginning of the nineteenth centuries. In the same vein Blanckaert's expression (2006, p. 48) to show the relatively recent nature of the disciplinary universe of disciplines is quite poetic and powerful (resonating as it does in a Borgesian way about the mythological founding of Buenos Aires): "[T]he disciplinary system was born yesterday. We, erroneously, believed it to be as eternal as the earth". He points out that the word itself did not explicitly appear until World War I.

According to Stichweh, the disciplines, like most inventions, were not the result of a moment of grace or enlightenment, or a singular lucky impulse, nor a mere institutional innovation. On the contrary, they were the result of many innovations over the course of at least six or seven decades. Stichweh (1992) describes the organization of modern disciplines on two levels: in internal terms, as a system of communication organized around several mechanisms, including their relationship with the professions and publications. In this sense they would seem to be relatively autonomous systems. But, on the other hand, he provides an overview of the entirety of scientific practices taken as a whole, emphasizing the dynamics of development and differentiation which generate the different disciplinary systems and within them, subdisciplines and sub-sub-disciplines. Blanckaert (2006) pointed out that the disciplines followed a dynamic of organized creation, or aligned with a project, with practical programs, and do not respond to a condition of access to the "nature" of things. Therefore, the distribution of disciplines is not based upon "natural relations" or logical classifications. We will return to this question later.

One of the classic and better presented examples is laid out in Ben-David and Collins' study on the origins of psychology somewhere between speculative philosophy and physiology. The authors challenge the idea that the emergence of a new disciplinary field is necessarily the result of intellectual growth within a preexisting disciplinary field, and instead propose placing emphasis on social factors to show that "a new scientific identity can precede and effectively make possible the growth of scientific production" (Ben-David & Collins 1966, p. 453).

As I suggested earlier, currently the debate about disciplines or other modes of organizing knowledge and the social group which investigates it seem to be neglected questions in the social studies of science. The general view is that the issue is already more or less "resolved" by studies over recent years and that it is better to develop an orientation towards observing the interactions between science and other social spaces, such as art, design, and urban living, as well as extending the social studies of science to other objects.[2] I consider that disregard as a mistake, not only because the issue is not truly "resolved", but the object itself—scientific practices and their organizational forms—is also undergoing a process of change and of formulating new challenges to analyze, understand, and interpret, as Marcovich and Shinn (2011) pointed out in one of the few studies that has continued problematizing this issue.[3] With this in mind, it is worth providing a brief overview of these different conceptions in order to be able to determine which concept and level of analysis is most suitable to give an account of our "beast" which is the development of science in Latin America.

Towards the end of the 1970s and the beginning of the 1980s, a few texts within the social studies of science focused on the question of disciplines, specialties, and other forms of approaching the dynamic of organizing the production of knowledge. In one of these texts Daryl Chubin proposed that

> with the institutionalization of science in universities the fragmentation of knowledge into intellectual provinces called disciplines was

legitimized. Disciplines, and the bureaucratic structures that support them, namely such as the academic departments, are charged with the training and certification of new scientists.

(Chubin 1976, p. 448)

In line with the mood of the time, in which everything related to the normative framework of functionalism was being challenged, Chubin points out that the most classic works of sociology had only dealt with some organizational or institutional aspects of knowledge, while the history of science was concerned with discoveries, ideas, instruments, and more generally the "conditions of progress" in science. They therefore constitute the well-known "black box" (Whitley 1972) which means that either the intellectual conditions and the content of knowledge or the social conditions under which these are produced are ignored.

Thus, for Chubin, the disciplines imply, above all, a fragmentation related to the training of new researchers, a crystallization of the distribution of scientific objects effective at socializing the young entrants to the social sphere of science. However, on the level of actual research practices, Chubin prefers to focus on a more specific space, that of scientific specialties, which he defines as "a viable concept whose various representations capture better than conventional units of analysis, especially 'discipline' the research processes and structure of research—and thereby explicitly ignores other scientific roles (e. g. teaching and administration)" (ibid. p. 449).

In the same year Chubin's book was published, the book edited by Lemaine and others about the emergence of scientific disciplines was issued. There the authors indicate that they are concerned with taking into account various aspects in the development of the disciplines that, until then, has been considered separately: internal intellectual processes, internal social processes, external intellectual processes, external social processes, the immediate institutional context, specific economic and political factors, and diffuse social influences (Lemaine et al. 1976, p. 14). As we will see, this represented a highly ambitious program which aimed at explaining practically all the existing dimensions in accordance with the varied lineage and backgrounds of the authors. Indeed, this collective book was the fruit of a Franco-British program called "Parex" (Paris and Sussex) which would go on to be the seed of the EASST (European Association of Studies of Science and Technology). In summary, it engaged with explaining the intellectual and cognitive factors in an articulated way, as was the trend at that time towards the end of the 1970s.[4]

The book published by PAREX tackles the emergence of various fields, from agricultural chemistry, physical chemistry, and radio astronomy, to tropical medicine, x-ray crystallography, and biophysics. They are, in some cases, extremely interesting cases and represent, to a certain extent, the foundations of the then-nascent field of sociology of knowledge; several of the book's authors would become indispensable touchstones of this field during the decades to follow: David Edge, Michael Mulkay, John Law, and Steve Woolgar. However, as Cambrosio and Keating (1983) rightly point out, despite the title suggesting otherwise, most of the book's articles are more focused on investigating scientific specialties than disciplines.

Just one year later Richard Whitley joined the debate with a similar concern to Chubin, that is, to give an account of the dynamics of investigation both in its intellectual and its organizational aspects. However, unlike the latter, Whitley advocated that the idea of discipline as a concept had sufficient analytical power, although his analysis is more subtle in the sense that he does not consider all disciplines to be equal. On the contrary, he states that "two important aspects of variation in scientific organization are the grade and type of specialization in the sciences. We can distinguish highly differentiated fields from those which are, comparatively, more homogeneous, and within a given range of differentiation, alternative organizations of social and cognitive structures" (Whitley 1976, p. 472).

Whitley attempts to put in order the considerations which seem to approach all fields of knowledge as necessarily equivalent and, in this sense, his analysis is very incisive. He proposes two types of disciplines: in the first type of disciplinary organization, which he terms "umbrella discipline", research production is predominantly organized at the specialty and research area level without direct reference to, or influence from, the discipline understood in broader terms. In the second type of organization, which he calls "polytheistic", metaphysical commitments which tend to focus on disciplinary identities can be observed (Whitley 1976, pp. 476–477). While the umbrella provides a very general view of science and a set of ideas for research which are concretized in specialties, the intellectual foundations of polytheistic disciplines are more strongly associated with current debates and issues. One of the reflections that Whitely makes in relation to the difference between these two disciplinary types consists of pointing out that, while in the polytheistic disciplines debates around the very definition of the discipline (or establishing its boundaries or legitimate objects, and so forth) are frequent, in the umbrella disciplines these struggles are less visible, or rather they take place within the specialties. Hence, the latter can compete among themselves about the relative importance—relevance, usefulness, originality—of each specialty within an overarching disciplinary

Cambrosio and Keating (1983) question Whitely over the fact that his analysis of the relations of authority in science is restricted to an interplay between disciplines and specialties, given that they are basically epistemic units of analysis. Instead, they suggest applying the concept of field developed by Bourdieu, which we shall return to later.

For most of the approaches mentioned thus far, despite their focus on the emergence and development of new disciplines or specialties, a kind of static character can be perceived insofar as the disciplines seem to emerge as a result of different mechanisms, such as the hybridization of preexisting fields, the disentangling or autonomization of a sub-discipline, or the convergence of new knowledge and practices at the diffuse boundaries of another field. However, once the emergence is underway, it is always problematic and disruptive since it reorganizes an already existing structure of social or socioinstitutional and episte-mological division. However, a new discipline generally does not remain crystallized as such, and a discipline with certain characteristics, both internal and in terms of its relationship with other knowledge structures, goes through a process of changes over time.

Terry Shinn proposed two approaches which could provide a useful contribution to this discussion. First, some years ago (Shinn 2000) he highlighted the existence of regimes of knowledge production, which he identified as "disciplinary", "transitory", and "research-technology communities" (RTC). According to Shinn, the first are relatively easy to identify, as they are rooted in clearly visible institutions such as universities and other research departments; they have distinct identifying markers and well-established regulations. In contrast, in the transitory regime its practitioners need to "step out of" their own field in order to search for methods, instruments, or theories from other more or less neighboring fields, although they then return to their discipline of origin. The third regime is independent of the disciplines since its practitioners lose their links to them and instead identify with a project, with specific developments not conducive to being framed within the existing disciplines. Of course, the transitory or RTC regimes may over time give rise to a new disciplinary field. However, this perspective, without overcoming the problems inherent to a disciplinary approach, contributes a highly useful dynamic element, approaching simultaneously the aspects of scientific practice and intellectual dimensions, as well as institutional organization issues.

Second, in a more recent study, Marcovich and Shinn (2011) return to the question of disciplines in order to refute those who claim the end or extinction of the disciplines and to formulate the characterization of a "new disciplinarity" arising, according to the authors, as a consequence of two main factors: the acceleration of the production of knowledge and the complexity. For them the new disciplinarity incorporates six elements: 1. Referent; 2. Research arena; 3. Combinatorials; 4. Projects; 5. Displacement, and 6. Temporality.

Marcovich and Shinn's proposal has two main strengths. On the one hand, it is one of the few within the social studies of science that problematizes the question of disciplinary organization and dynamics, while most scholars seem to disregard it as though it were "already resolved".

This stance hinders giving an account of the transformations which have occurred over the last few years, when a significant part of the disciplinary organization was disrupted by the winds of change. On the other hand, Marcovich and Shinn elaborate a very keen analysis in their study of historical dimensions, especially from the establishment of the "traditional" disciplines during the nineteenth century (astronomy, organic and inorganic chemistry, physics, biology, mathematics). However, in a decision which we consider regrettable, the authors decided to limit their analysis to the processes of cognition and epistemology which "constitute the basic unit of analysis", and thus explicitly exclude any question related to contextual, social, or economic factors. As they assert, it is "imperative to remember that cognition comprises the primary intention and expectation in science" (Marcovich & Shinn 2011, p. 583). Wishing to be explicit and to avoid the reader confusing them with any of the currents that speak about "interdisciplinary" perspective, they add: "By cognition we refer to the basis on which knowledge is developed; this includes questions-formulation, methodology, practices of exploration and evaluation criteria". Disregarding all noncognitive aspects is a sad decision by the authors because it means that while, on the one hand, their approach represents a crucial contribution to returning to debate the discipline as one of the organizational axes of contemporary science, on the other hand, it represents a huge step backwards in the study of sciences.

As mentioned earlier, Cambrosio and Keating suggest working with an adaptation of the—now well-known—concept of scientific field advanced by Bourdieu. I will provide a brief summary for readers who are unfamiliar with the idea. For Bourdieu a field is a space of struggle in which practitioners compete to obtain the greatest symbolic capital (*scientific* capital in the case of science), which enables them to wield the power to establish the limits of the field and define what is and is not legitimate, as well as the ability to define the ways in which that capital accumulates. The exercise of authority requires recognition by others (peers-competitors) of scientific competence and skills (cognitive authority) and the ability to impose the rules of the game on the field (political authority). Therefore, every field is organized according to differential endowments of authority, sustained by an unequal distribution of scientific capital which organizes it into dominant and dominated, all functioning in a space which is relatively autonomous in relation to other social spheres.

Without a doubt, Bourdieu's approach proved to be of great value to break with the idealized version dominant until the 1970s of a domain of science understood as a community regulated by consensual norms and in which cooperation and solidarity prevailed. It is also a sharp tool for analysis, inasmuch as for this approach technical and even epistemic definitions are not untethered from the ensemble of social relations and, above all, power, but are seen as simply two sides of the same coin (Knorr

Cetina 1999). For this reason, it has been tempting to adopt the concept of field for the analysis of diverse scientific spaces, and this has been the case in Latin America above all, where a generally acritical application of these categories has been carried out (Kreimer & Vessuri 2017; Kreimer, Vessuri, Velho, & Arellano 2014).

From my point of view, however, this approach has various problems. The first, already highlighted by Knorr Cetina (1983) many years ago, concerns the analogy with a "quasi-economic" model in which one type of rationality prevails over any other: In effect, the practitioners in a field seem to be driven by the logic of the accumulation of capital above and beyond any other motivation of a cultural, affective, sociocultural or even aesthetic order. Abundant empirical studies have shown the multidimensionality of scientific work, similar in this sense to any other social practice and, therefore, not reducible to a single type of rationality. Second, Knorr also calls attention to the fallacy of autonomy, which in Bourdieu and also in Merton's prior studies had not only a descriptive but also a normative character. In other words, science not only is perceived as autonomous (from all political, economic, religious, and other influence) but it *must be* autonomous.[5] Thus, if a scientific field were really autonomous, the question of resources would generate an immediate contradiction (as Knorr astutely suggests), since the assumption would be that "someone" would (or should) supply the resources needed for the field to function, while at the same time limiting all intervention from that source. The history of science refutes that narrative.

An additional problem which the concept of the field as proposed by Bourdieu poses is the correspondence or relationship between field and discipline (or specialty) and especially the relationship between different disciplines, their boundaries, and their frontier zones. Lenoir (1993, 1997) partly borrows from Bourdieu (and Foucault) as a basis for his approach to the study of disciplines. For Lenoir, disciplines are "dynamic structures for assembling, channeling, and replicating the social and technical practices essential to the functioning of the political economy and the system of power relations that actualize it" (Lenoir 1993, p. 72). According to him what we can observe, instead of "monolithic disciplines", are "disciplinary programs locally adapted to the political economy". The disciplinary programs are instruments for defining society by organizing and "packaging economies of practices for specific clientele" (Lenoir 1993, p. 73). As institutions which demarcate "the limits of expertise and hierarchy of competence, disciplinary programs are generated simultaneously within political and ideological discourse and therefore best be understood as a discourse of power as well as instruments of knowledge production" (Lenoir 1993, p. 86).

This approach is interesting to the extent that it incorporates the inextricable character of the generation of knowledge—and discourse— legitimated in a social context and the function of authority as part of the same mechanism. Although Lenoir does not clarify how to deal with some of the problems we indicated in relation to Bourdieu's approach, particularly the notion of autonomy and the relationship between different disciplinary spaces, he does tend to assume that each discipline forms a specific field—a field of knowledge production, a field of dispute, a field of legitimation, and a field of power—each of which struggles to establish itself. This occurs in the case of *successful disciplinary programs*, which he contrasts with *research programs*, which, guided instead by a particular problem, do not aim to generate an institutional and institutionalized space.

Stichweh puts forward something similar, questioning why disciplines still exist if, as many seem to believe, most scientific activity takes place within sub-disciplines or sub-sub-disciplines. His response is largely based on two issues: on the one hand, the fact that many disciplines provide a general perspective which goes beyond the sub-disciplines, which consequently enables internal mobility of occupational roles within its remit; on the other hand—and here he draws closer to Chubin—there are the strong roots they have in educational structures, from schools to universities (Stichweh 1996, pp. 13–14).

To conclude this section we should, at least briefly, mention two important issues related to the formation of disciplinary fields: the role of leaders and that of the technical dimensions of research.

On the role of leaders, an issue which Lenoir also engaged with although did not explicitly problematize, it is worth stressing that it is not always the case that the initiators of a new research area are eventually considered to be the "father" of a new disciplinary field. Indeed, the results cannot be reduced to the strategies of the actors at a given moment, nor to their technical skills or political endowments, but rather to a set of factors which also includes institutional dynamics (which may be more or less rigid according to contexts and time periods), the more or less firmly established character of the preexisting disciplines, the given conditions for the development of scientific careers, and the availability of resources, among other significant variables. To paraphrase Latour (1989), it is not until after a new disciplinary field has succeeded in its institutionalization that its boundaries, objects of study, methods, instruments, and theories seem to be evident, natural-seeming, and in the last instance, are "blackboxed".[6] This means they are part of a black box which will not be subjected to questioning for a considerable variable (usually long) period of time. But the causality is never reversed: Given a set of problems to study, a collective of actors, theories, and instruments will naturally take shape to investigate and explain it. But rather we argue here exactly the opposite: This object of study is coproduced at the same time as a social, instrumental, cognitive, and institutional group becomes established and is stabilized.

In relation to the role of leading figures in the emergence of new fields, Gustin (1973) directed his attention to these "charismatic" personalities in science, without which individual or collective ventures would often fail. This is interesting as it complements and challenges the idea of "structural factors" which explain the birth of new specialties and fields. If, on the contrary, as we have been arguing here, these dynamics are highly contingent, understanding the role of these leaders is a crucial task. Ben-David and Collins (1966) in their aforementioned pioneering study had already documented the role of the "founder" in the origins of psychology, while Mullins (1972), in his famous article, strongly influenced by Kuhn, about the phage group and the origins of molecular biology, described these figures as "intellectual leaders" whom he distinguished from a concomitant role which he called the "organizational leader". According to Mullins both are necessary for a specialty or discipline to be established. Along the same lines Shinn (1988), in a strongly Bourdieusian text, claims that the hierarchies within a field do not comprise an indivisible whole, but rather they are composed of two dimensions: social hierarchy and cognitive hierarchy. However, unlike Mullins' approach, Shinn considers that in most cases both tend to coincide in the same personalities; those located on the highest rungs of the social hierarchy within a field are simultaneously those who enjoy the highest positions at the top of the cognitive hierarchies. However, when a discrepancy arises, for example, when we are faced with someone with an elevated position in the social hierarchy but with a low cognitive ranking (or vice versa), there is a very high probability that serious conflicts will emerge.

The role of instruments is too complex to properly deal with on these pages. One useful point would be to return to Whitley's suggestion of differentiating types of disciplinary fields and analyzing to what extent instruments can determine the formation of a new area autonomous of the rest. In addition, Hacking's (1992) conceptualization differentiates "laboratory" sciences ("self-vindicated" through the use of their own instruments) from others, a distinction which could be useful initially but which is a little static when certain practices leave and/or are brought into the laboratory over the course of a field's history.

The analysis of the role of instruments and, more generally, the role of technical resources in scientific research and its modes—both social and epistemic—of organization over the course of the history of science offers us rich examples of varied configurations, and the construction of a typology of the role of technical dimensions in the emergence of new disciplinary fields would be a considerable task (but it is evidently one which we cannot undertake here). Let us briefly take a look at a few of these configurations, just to provide a glimpse of the type of analysis possible. One of them is, without doubt, that of a disciplinary field emerging almost directly from or closely connected to a new technique or device, such as the case of crystallography, developed at the beginning of the twentieth century by the Braggs (father and son). They began to study crystal structure thanks to x-ray diffraction, which basically constituted a technique of visualization of structures (Burke 1966). Thereafter the use of this technique has been used for crystallography of metals, playing an important role in the subsequent development of materials science (Bensaude-Vincent 2001), or protein crystallography, in which John Bernal would be a pioneer and would be intimately associated with biochemistry and later with the birth of molecular biology (Stent 1968; Morange 1994). Thus, for this first type, a technique is at the heart of the birth of a field which immediately branches off in two quite separate directions.

Another well-known case is nuclear physics, which was highly dependent for its development on the building of a particle accelerator which could break the so-called "Coulomb barrier". The first of these, called the cyclotron, was constructed by Ernest Lawrence at Berkeley, practically at the same time that the neutron was discovered (Reed 2014; Pestre 1992). Both developments, occurring between 1931 and 1932, would be key to the Manhattan Project, the building of nuclear bombs, and the subsequent evolution of this field.

* * *

Certainly, the most suitable level of analysis is composed of a combination of perspectives, some situated in what could be described as a "specialty" (in Chubin's sense), others in broader fields closer to Whitley's concept of the "umbrella", while others might be studied in their dynamics as spaces or regimes in transition, whether due to hybridization or differentiation. Each conceptual definition should be operationalized on the basis of what is being problematized or highlighted at particular historical moments. The subtext underlying this decision is the notion that analytical richness rests on the possibility of analyzing diverse spaces for the production of knowledge, communication, sociability, organization, and so on. Indeed, limiting ourselves to an unduly narrow definition would imply losing much useful information.

In general terms, however, one could adopt a certain concept of the field in which subjects behave according to their own interests and in search of symbolic capital which would enable them to influence the definition and boundaries of their own field. But this approach moves away from the associated analytical consequences the concept brings with it, such as the existence of more or less clear and established boundaries (inside and outside of science), autonomy and the lack of participation by other relevant actors in the dynamic of the field, or the rationality of accumulation as the only or dominant factor which explains the subjects' practices. Instead, we adopt a concept of a field which has defined but highly porous boundaries with a relative and provisional stability, in which the cultural representations and the interests of the agents go beyond the search for greater authority, whether social or cognitive.

Another crucial point at which we seek to distance from a Bourdieu-sian perspective of autonomous fields is in the relationship between these spaces and the society they are embedded in. Indeed, if Bourdieu (1997) proposes that we reject as falsely constructed the question of the social demand for knowledge, as according to him it seemingly hides a euphemism for the appropriation of public knowledge (financed by public funds) by capitalist firms, we consider that the question, far from being resolved, is problematic. Bourdieu addressed this text in his talk to the researchers of the French National Institute for Agricultural Research (INRA), and he advises, even exhorts, them to pay no attention to such demands (that actually come from the science policy makers who represent business interests) and instead focus on accumulating symbolic capital within their field in order to be able to intervene more successfully in the public sphere.

To put this in context: In the France Bourdieu writes in, industry accounts for more than half of total expenditure on research and development (R&D), while the average for OECD countries (the most economically advanced) is around 60 to 65 percent, depending on the year (www.stat.OECD.org). By contrast, for Latin American countries, although with unreliable indicators, this percentage rarely reaches above 20 percent of the total. This means that science and technology in the central countries are, for a large part, financed by and oriented towards the interests of industrial development. So, Bourdieu is addressing a group that represents no more than 40 percent of knowledge nationally produced.

This issue has a rather different sense in Latin America, where the percentage of scientific knowledge effectively used to tackle social problems—of health, production, or environment—is tiny (Kreimer 2015; Kreimer & Zabala 2009; Kreimer & Thomas 2005). Therefore, the dimension of the utility of knowledge would be an important variable to take into consideration when analyzing the emergence of new scientific fields in Latin America in a much more crucial way than for the international scenario, as we shall show in the next section. Additionally, the imitative character of some of the initiatives developed in institutional design, and also in the orientation of policy agendas and the practitioners in each field, is an important factor which should be mainly considered in the analysis.

Scientific Fields in Nonhegemonic Contexts: Do Centers and Peripheries Still Exist?

Stichweh, in one of his articles (1992, p. 4), suggests that he is not concerned with studying "the preconditions for the establishment of a scientific discipline in the context of other, previously established disciplines", but rather "the preconditions for the establishment of scientific disciplines per se, at a moment in history when it is unknown that a representative universal social form (the scientific discipline) is coming into being". Naturally, he does not state that he is going to speak about England, France, and Germany, since that is taken for granted. We encounter the same assumption in most of the renowned scholars in the social, historical, or political studies of science. Just as in the Anglo-Saxon world the term "science" does not include the social sciences, but merely the "sciences", likewise analyses about the dynamics of *knowledge production, scientific debates, laboratory life, trans-epistemic arenas, symbolic fields of production, obligatory passage points, relevant social groups*, and many other concepts, do not need to clarify that they refer to the developed world.[7]

When we refer to "Western science", meaning the science which seems to have emerged and been institutionalized in England from the seventeenth century onwards (Merton 1938), we are speaking, naturally and with no need for clarification, about the science, institutions, practices, and actors from the developed capitalist countries. Furthermore, this object requires no justification: to study the origins of the French literary field (as Bourdieu does in his book on Flaubert) is to study the literary field *tout court*. As MacLeod (1980) states, "The Transit of Venus expeditions were thus 'metropolitan', not just because they were launched from London or Paris, but because they implied a set of intellectual structures and questions common to the Metropolis. *Metropolitan science was science*" (MacLeod 1980, p. 2; my emphasis).

This seems natural for authors for whom the locus of knowledge and its dynamics is not problematic (whether due to their sole focus on intellectual or theoretical development, or because they conceive of science as a necessarily universal practice). However, it is surprising that—with very few exceptions—scholars in the sociology of scientific knowledge have also naturalized the fact that the object of all their research is that which we could call *central* or *mainstream science*. Furthermore, until recently, that does not strike them as being problematic.

On the contrary, when studying the development of science, scientific debates, laboratory life, or the dynamic of knowledge production in Latin America, Asia, or Africa, it is imperative, above all, to justify the legitimacy of speaking about "science" in those contexts and, following that, to explain that the studies analyze some particularly interesting case which could shed light on some highly relevant question or unique feature, or that it will dazzlingly illustrate some process of considerable interest. To put it another way: Said object (science in those countries or regions) is not *interesting per se*, as "Western science" is, but its interest must be constructed.

At least on this point the much-discussed concept of symmetry, originally advanced by Bloor in 1976 and widely discussed by authors such as Harry Collins and Bruno Latour, among others, seems not to hold water: No longer can we speak of symmetry in the analysis of true or false premises, or the extended symmetry between the natural and social worlds. Indeed, when analyzing the diverse contexts of science and its practices, certain contexts require justification and others do not.[8] It remains ironic that when Latour justifies his choice of the Salk laboratory/a laboratory at the Salk institute for his—later—famous study of laboratory life, he simply comments that

> The choice of laboratory was determined mainly by the generosity of one of the senior members of the institute in providing office space, free access to most discussions and to all the archives, papers and other documents of the laboratory, and part-time employment as a technician in the laboratory.
>
> (Latour & Woolgar 1979, p. 39)

Of course, the roots of most scientific fields, disciplines, and specialties can be found in the most advanced countries where, in spite of some interesting transformations over the last few decades, the greater part of global scientific production has been, and still is, concentrated. Indeed, until twenty years ago 37 percent of the global production of articles in indexed journals originated from the United States, and 35 percent were coming from the European Union, meaning that both regions together accounted for three-quarters of the total papers production. If we add four more countries to this panorama (Canada, Japan, Russia, and Australia) which together represents a little more than 20 percent, the figure reached 93 percent of the total, so that all the other countries shared the remaining 7 percent (OECD [n.d.]). Currently this outlook has been slightly modified: The production of the United States has dropped to 27 percent, the European Union has maintained its share to 35 percent (mainly due to the increase in the number of countries: from fifteen to twenty-eight), and if we add the production of the same four countries mentioned earlier (12 percent), we now arrive at 74 percent, meaning that the remaining countries share a quarter between them. However, half of that is now taken up by China and India as new powerful centers of knowledge production (Levin, Jensen, & Kreimer 2016).[9]

Over time many studies have attempted to explain scientific development in relatively less developed countries, beginning with George Basalla's classic text (1967) on the spread of Western science. There he advanced the premise that there is a type of "single path" in scientific development; there are three stages which the "non-European" nations must follow for the introduction of modern science. Basalla chooses some verifiable cases and expands from there an analytical model: The first phase is characterized by a "nonscientific" society—located in emerging nations—which European scientists visit in order to take away natural history samples, astronomical data, and so forth. The second phase corresponds to "colonial science" and is marked by dependence on the institutions and traditions of the center. The third phase is when modern science takes root in independent scientific traditions, entailing changes in order to overcome "resistance to science on the basis of philosophical or religious beliefs".

It is remarkable that this apparently single track parallels the stages of economic development proposed by Rostow (1962) only a few years previously. He had proposed five stages instead of three, but the guiding logic is quite similar: There is only one path of economic growth (development), and it is more or less the same for all countries. For Rostow, the first stage is that of traditional societies, and in the second the conditions for "take-off", which is the key concept in his analysis, are established, and the demand for the export of raw materials increase, infrastructure is built, and the social structure begins to be modified. The following stage constitutes take-off itself, during which the process of urbanization is completed and the manufacturing sector is developed. The fourth stage is the drive to maturity, in which the manufacturing base is diversified, communications and transport infrastructure are expanded, and there is investment on a large scale. Finally, the fifth stage corresponds to mass consumption as a result of the industrial base dominating the economy, and the consumption of high-value manufactured goods (such as cars) becomes widespread.

Both models have been subject to rigorous criticism. Regarding challenges to Rostow's model (imbued with a certain common sense of the era along with a large part of the conceptual basis of the policies), the ideas of Prebisch and the Economic Commission for Latin America (ECLAC) around the hypothesis of the long-term deterioration in the terms of trade for primary goods and food, on the one hand, and manufactured goods, on the other, were a contribution of great importance. According to Prebisch (1951, p. 3) "the universal propagation of technical progress from the original countries to the rest of the world has been relatively slow and irregular". He adds, "[I]t is not possible to understand the problems of economic development in Latin America without examining that process and its consequences" (ibid., p. 48). By the 1960s several authors, such as Gunder Frank, had advanced much further in the demystification of "a single path" of development, proposing, for example, that "the historic development of the capitalist system has generated underdevelopment in the peripheral satellites whose economic surplus was expropriated while producing development in the metropolitan centers which appropriate that surplus" (Gunder Frank 1965, p. 12). Thus, instead of stages or phases along the same path, development and underdevelopment are perceived as "two sides of the same coin". We can find analyses close to that of Gunder Frank in a generation of critical Latin American authors, such as Pablo González Casanova (1969), Ruy Mauro Marini (1973), and possibly the widely known book by de Fernando Henrique Cardoso y Enzo Faletto, "Dependency and Development in Latin America" (1969).

Outside of Latin America, one of the studies with the greatest impact at the time was Samir Amin's (1973) study of unequal development. For Amin, "the genesis of central capitalism constitutes the first great expression of the law of unequal development of formations".[10]

The scientific critique of the development of science in different contexts moved in different directions. First, beyond Basalla's model, we find a number of studies which analyzed the "reception" of (Western or European) science in underdeveloped contexts. These studies about reception can be divided into two disparate types of analyses: those which emphasize the conditions of transmission and aspects of an institutional type, such as the creation of new institutes (more or less similar to those existing in the metropolis).

From this perspective, local society seems to play the role of an empty vessel prepared to receive what comes from abroad, while local actors play a somewhat neutral role, being fairly receptive to new theories, methods, disciplines, or institutional designs. These approaches often fall into a dualism, already insinuated by Basalla, of positing an opposition between the modern (European science) and the ancient (local beliefs).

One variant of this type of analysis is the familiar exegetic or hagiographic type of history of science, pointing out the specific and exceptional conditions of the "local pioneers" of science, who usually reproduce "modern science". Certainly, this type of approach which entails the accumulation of accounts of the greatness and genius of providential men like Newton, Darwin, and Laviosier (and many more) found less adherence in science studies in the underdeveloped countries, since the emergence of these "heroes" is less frequent in peripheral contexts, although even in these contexts there has been a strong temptation to find their own providential men.[11] In fact, this corresponds with a way of writing history, a school of historiography which focused (and in some cases still focuses) its attention on individuals, their qualities, and virtues. Most of the history of humanity has been told or reconstructed in this way, at least until well into the twentieth century. A common feature in this type of history of science, both in its "rationality versus barbarism" version and the kind which focuses on the great pioneer, a kind of romantic hero, is the duality that it implies: A society is judged *de facto*, to be backward or subdeveloped, categories which only acquire relative meaning in comparison with Western European ideals of civilization. It should be noted—in order to appease the burden placed on historians and their predecessors—that the history of science, unlike other branches of history, is frequently practiced not by professional historians—understood as those who have some specific background in historic research—but by active or (some prematurely) retired scientists who embark on the venture with a reconstruction of their own discipline. While it is less common for priests to write the history of religion or marginalized populations to write the history of social policies, the fact that scientists become historians of science occurs more often, and not always to the benefit of this history.

Of course, in all these histories science is presented as something positive, as apolitical and neutral in its values, and therefore its spread implies assistance, free of any particular values, for material progress and civilization (MacLeod 1980). The idea underlying studies of the "spread" or diffusion of European science is that, definitively, science is a universal practice and one that is profoundly rational and independent of the context in which it is practiced. Hence, the only obstacles to overcome are those irrational beliefs which hinder its development in the search for the construction of objective truths. From this standpoint, modern (Western or European) science is one of the manifestations of human progress, a provider of well-being, and a helping hand in the creation of wealth for the whole of society.

As we noted earlier, the analyses which take this approach to investigating development or the implantation of science in underdeveloped countries sooner or later bump into the struggle between two logics: that based on the concept of progress rooted in modern science, and others that are not yet capable of understanding the methods and theories with which to investigate and rationally understand the enigmas of the physical and natural world—that is to say, in order to universally establish the rules and laws which govern them and that, therefore, society could use for its benefit as occurred first of all in England, beginning in the seventeenth century, followed by other European societies like France, Germany, the Netherlands, and so on.

A second line of analysis about "reception", developed in the 1960s and 1970s, adopts a more critical view, analyzing the cultural conditions of the local societies, the pattern of conflicts, representations, ideas, and the like. Without wishing to launch into a full engagement with a considerable number of studies, it is unavoidable that we refer to the writings of Lewis Pyenson, who studied in great detail the process of the expansion of European science across various texts. In *Cultural Imperialism and Exact Sciences* (1985), Pyenson examines the first years of the twentieth century, when German physicists and astronomers established themselves in research institutes and universities in Argentina, the South Pacific, and China over a period of three decades. We should bear in mind that, until the 1930s, German physics was remarkably dynamic, claiming nearly half of the Nobel Prize winners (the Nobel Prize first being awarded in 1901). Pyenson's thesis suggested that, despite the exact sciences being of relatively little practical utility, they represented an important tool of cultural imperialism. This was a strongly debated claim in the sense that Argentina, in those years, "was the second most important place for theoretical physics after Germany" (Ciencia Hoy 2001). In another of his detailed studies (with an impressive degree of documentary evidence) Pyenson (1993) addressed French science, reviewing the work of French scientists in Algeria, Morocco, Indochina, China, Lebanon, Madagascar, Chile, Argentina, Brazil, Mexico, Cuba, and Martinique. Taking the differentiation between "exact sciences" and "descriptive sciences" as a starting point (just like his book about German science, but in a more pronounced way here), one of his main arguments is the following: "given that the exact sciences resist ideological contamination, they should serve as a test to study how knowledge sustains the expansion of political power". Naturally this strict, and ultimately artificial, distinction between different fields has been challenged by Dear (1994), Headrick (1995), and Harrison (1995), to name a few. Above all, this distinction does not allow us to analyze the complex interactions between disciplinary fields—such as medicine, engineering, and agriculture—or the use of knowledge applied to varied problems related both to local and to European societies. We will return to this important issue later.

However, the most debatable aspect of Pyenson's work is his scant attention to local communities and cultures, as if he implicitly assumed the values of the French or German scientists for whom everything that differed from their hegemonic perspective represented a barrier to the development of "good science"—or just to science full-stop. The author's defense, alluding as it does to the "scarcity of sources", is, in this sense, unconvincing.

Conversely, Vessuri (1994c, pp. 184–186) points out that the introduction of "Western" knowledge spawned a host of reactions that slanted its assimilation or rejection. She stresses that it makes little sense to consider "colonial" societies as homogeneous or immutable spaces; rather, "they have not stopped defining and redefining themselves". Less developed societies have been confronted with a constant dilemma: In some sense, they are perceived as similar to (modern) metropolises, but in another sense, they are also quite different from the most developed countries (or, in colonial terms, "countries of origin"). Indeed, ambiguities or ambivalences have accompanied the process of institutionalization of science in colonial countries, where "the modern", which deals apparently effectively with a variety of issues, coexists—often in conflict—with traditional perspectives, which are highly dependent on specific contexts and social configurations.

We can now look at four issues that have altered the perspectives on the study of science in developing countries. This operation, at least in part, occurred in parallel with the questions on the theory of development discussed earlier: First, we have the emergence of the "center and periphery" approach, from its original proposal by Edward Shills in 1961 and on through all its subsequent developments. Second, there has been the rise of the sociology of scientific knowledge (SSK), or the constructivist metaphor (Sismondo 2004), since the mid-1970s. Third, there is the deployment of studies on colonial and imperial science, viewed from the perspective of the process of development, hybridization, conflicts, or tensions in each developing context. These studies replaced the previous perspective, which emphasized the expansion or "diffusion" of central science. Finally, the relatively recent trends are grouped under the label "postcolonial studies".

These currents are usually analyzed separately, partly due to the disciplinary confinement where historical studies on science, political, and cultural studies and sociological studies on scientific practices have been parceled. I believe, however, that we must consider them as different facets of a prism: These approaches entail breaking—analytically and by various other means—with the traditional notion of science as universal, neutral, objective, progressive, and accessible to all those wanting to produce knowledge. Naturally, this is strongly associated with the ideal of modernity (at least, to the ideal of modernity held by the elites of the most dynamic countries in the developing world).

Studies on the sociology of scientific knowledge, as well as historical studies, especially the so-called "new history of science", which arose simultaneously,[12] undoubtedly made a fundamental contribution in breaking with idyllic and rationalistic images of hitherto prevailing scientific practices. In fact, authors like Collins (1981b, 1983) stigmatized these traditional approaches with the acronym "TRASP": knowledge was perceived as True, Rational, Successful, and Progressive. In contrast, these studies showed the weight of the "locality" through new methods and theories: entering into spaces where knowledge is effectively produced and analyzing the complex negotiations that take place, both at the discursive level and in the mobilization of the natural world to show its artificial, constructed nature. Indeed, they showed how social, economic, cultural, and affective factors play a fundamental role in the technical and cognitive processes of knowledge production.

However, almost all the studies within the sociology of scientific knowledge consider, in a *natural* way, the science done in hegemonic countries. As Rodríguez Medina (2014) has pointed out, both past and current studies belonging to constructivist approaches offer insightful studies on laboratories, universities, and research centers located in the metropolises, but the highly precise descriptions of these phenomena cannot be transferred directly to peripheral contexts.[13] We will come back to this question in the last section.

Therefore, for many years, the research conducted from the mainstream of STS studies has focused on *mainstream science* with the explicit aim of showing its local character, its production processes, its negotiations, and so forth. As I pointed out in the previous chapter, they did not even bother to draw attention to the fact that they were studying "Western science", the way dissemination studies did, nor "science of the North" (as it began to be called a few years later): To study English, French, German, American, or Dutch science meant to study "science" period.

Let us go back now to Shils's (1961, 1975) classic text, somewhat forgotten nowadays. It is worth quoting an extract (extended as it is), when he posits that:

> Society has a centre. There is a central zone in the structure of society. [. . .] The central zone is not, as *such*, a spatially located phenomenon. It almost always has a more or less definite location within the bounded territory in which the society lives. Its centrality has, however, nothing to do with geometry and little with geography.
>
> The centre, or the central zone, is a phenomenon of the realm of values and beliefs. It is the centre of the order of symbols, of values and beliefs, which govern the society. It is the centre because it is the ultimate and irreducible [. . .]
>
> The centre is also a phenomenon of the realm of action. It is a structure of activities, of roles and persons, within the network of institutions. It is in these roles that the values and beliefs which are central are embodied and propounded.
>
> The mass of the population in most pre-modern and non-Western societies have in a sense lived *outside* society and have not felt their remoteness from the centre to be a perpetual injury to themselves. Their low position in the hierarchy of authority has been injurious to them, and the consequent alienation has been accentuated by their remoteness from the central value system.

He concludes with a pinch of optimism:

> Nonetheless, the expansion of individuality attendant on the growth of individual freedom and opportunity, and the greater density of communications have contributed greatly to narrowing the range of inequality. The peak at the centre is no longer so high, the periphery is no longer so distant.

This perspective inspired several texts that criticized diffusionist models and, especially, universalist ones, and advocated the existence of center and periphery in the universal scientific sphere. This occurred in parallel with the earlier-mentioned questioning of economic development, as well as the rise of dependency theory and other critical perspectives. Shils emphasizes, for example, tension and ambivalence, especially in the stance of intellectuals who are spatially located in the periphery but whose mental maps are, we would say today, "formatted" by the metropolis. According to him, these "cosmopolitans" are fundamental agents of change, so they have the ability to steer other colleagues (Inkster 1985). This type of statement is clearly in line with authors like Oscar Varsavsky, well known in Latin America (and almost unknown abroad), who has, among many other topics, stated the problem of what he calls "cultural dependency":

> It is natural [. . .] for young scientists to look with reverence to that Northern Mecca, believing that any direction indicated there is progressive and unique, going to their temples to improve themselves, and once they have received their support, returning—if they do return—while keeping a stronger bond with that Mecca than with their social environments. Then they choose one of the topics in vogue there and believe that this is *freedom of research*, just the way some believe that being able to choose among six newspapers is freedom of the press.
>
> (Varsavsky 1969, p. 15)

Varsavsky's idea about "scientificists"[14] is very close to Shils's notion of "cosmopolitans": The social role is similar, as they are the mouthpieces and symbolic operators within peripheral societies of the values and interests of "international science" or "central societies". We see here, therefore, a true "sign of the times": Center and Periphery, or North and South, were proposed then as binary poles of the unequal distribution of resources, capacities, devices, and, in general, symbolic and material goods. The studies conducted in the 1960s and 1970s were grounded in a strong structural perspective: The "peripheral condition" in scientific terms was anchored in the structurally peripheral societies marginally established in the world system. Moreover, it was reproduced by the local elites, who willingly assumed the role of active agents in the alignment of peripheral science with the dictates of "central" or advanced science.

Lafuente and Sala Catalá (1992) proposed an analytically fine and, therefore, interesting approach to this question. They suggested three criteria to analyze emerging concepts and approaches: geopolitical, socioeconomic, and socioprofessional. According to them, concepts such as dependent science, national science, marginal science, and academic science will vary depending on where the emphasis is placed. Thus, all these concepts involve the challenge of imagining modern science according to its weak, derivative, disjointed, or inferior forms as compared to the institutionalization of science "in the West".

These dimensions provide far more clarity than the sharp, binary definitions widely circulated for several decades, while also allowing different viewpoints to observe these contrasts, not limited only to the opposition of modern versus traditional or backward versus archaic. According to Lafuente and Catalá, from the geopolitical point of view, there is a clear and specific distinction between peripheral and metropolitan science: While the former refers to institutional mechanisms that function as enclaves in an economically less powerful country, the latter consists rather of expeditions of metropolitan scientists—agents of colonial powers—in the colonies to accumulate data (mainly in fields like botany, astronomy, and medicine). Data will then be processed in their country of origin and capitalized by their institutional and political structures (Lafuente & Catalá 1992, p. 16).

Lafuente and Catalá state that the emphasis on geopolitical conditioning has displaced the center of gravity of the more antiquated studies; they follow MacLeod (1980) in no longer considering science *in* imperial history, but science *as* imperial history. They rightly observe that

> the institutionalization of a colonial emplacement of a given imperial policy not only accounts for a process of expansion in the center-periphery direction, but simultaneously establishes the limits on expansion operating in the opposite direction, thus modifying the metropolitan policies; globalist and localist positions are, therefore, two sides of the same coin.
>
> (Lafuente & Catalá 1992, p. 17)

From a relatively close perspective, and as a critique of Basalla's diffusionist approach, Roy MacLeod delved more deeply into the very relationship between center and peripheries. He proposes a more dynamic conception of so-called "imperial science": He distinguishes heterogeneities and different types of relations within different countries and regions, as, for example, between what he calls the "occupied empire" of India and the "informal empire" of Latin America.[15] More importantly, according to MacLeod, there is no linear extrapolation of ideas between empire and other contexts, but rather multiple autochthonous developments that have "reverberant effects". Instead of "enlightened" metropolises radiating out, there are "moving" metropolises, as a function of the empire that selects and cultivates intellectual and economic frontiers. He points out that "it was the peculiar genius of the British Empire to assimilate ideas from the periphery, to stimulate loyalty within the imperial community without sacrificing either its leadership or its following" (1980, p. 14). In other words, not only valuable knowledge was produced in the peripheries, but that knowledge could be and was exploited by the metropolis.[16]

From a rather similar perspective to MacLeod's, the historian Kapil Raj also tries to overcome both the diffusionist models (Basalla style) and the confrontationists, who see the developed world or the metropolis and the peripheral world as two antagonistic binary pairs. This is an important step in understanding the development of science in nonhegemonic regions. From this perspective, and analyzing the scientific development in seventeenth-century India in relation to the British Empire, he shows, on the one hand, the complexity of interactions in the construction of modern science, even in an "asymmetric colonial situation" and, on the other, the active role played by heterogeneous knowledge networks in forging British identity, not to mention its research and training traditions (Raj 2000, p. 133).

The text by Vessuri (1994c) is along the same lines, pointing out that there are two sides to science in developing countries: On the one hand, it expresses the interests of the more advanced countries, but also the vigorous efforts deployed in the Third World to dominate and make productive use of scientific knowledge that bears promises of modernity. She concludes by pointing out that the creation and development of scientific institutions in developing countries is a necessary, but not sufficient, condition for success or failure in identifying problems and finding solutions.

A new type of analysis was emerging, much more focused on the type and content of interactions occurring in peripheral territories—albeit not limited to them—among the different cultures in play and, particularly, between heterogeneous arenas where structural dimensions are significant but do not define before the fact the whole set of real and symbolic dimensions that arise in knowledge production processes. In a recent text, Raj (2016) points out that it is necessary to pay attention to new forms of "relational" history that highlight the intermediations or "go-betweens". This should allow us to observe the various translations taking place and the more complex production of knowledge as a product of cultural encounters and conflicts, cooperation, and competition as affected by interests and needs, and so on.

The question of assembling the locality of knowledge (which includes practices, cultures, languages, and territories) and the dimensions linked to the international location, including non-egemonic contexts, is a task that will take several years, as we will see towards the end of this text.

Postcolonial Studies: When the North Discovers That "the South Also Exists"[17]

So-called postcolonial studies, which have emerged in the last two decades, question the impact of the colonial structure on the economic, political, social, and epistemic conditions of the colonies (Rodríguez Medina 2014). According to some of the authors adhering to this perspective, the practice of various colonial actions has had, as a result, the transmission (from Europe) not only of a set of institutions and social relations but also an ordering of knowledge hierarchically structured in previously established disciplines as an epistemic distribution of the world. Postcolonial studies has paid attention to the epistemic foundations of knowledge production and found that the voices and communities of those countries were repressed and not taken into account by metropolitan science, giving rise to a phenomenon described as the "coloniality of knowledge" (Rodríguez Medina 2014). Mignolo, one of the most representative thinkers of this school, points out that

> The main thrust of my argument has been to highlight the colonial difference, first as a consequence of the coloniality of power (in the

making of it) and second as an epistemic location beyond right and left as articulated in the second modernity [. . .] The world became unthinkable beyond European (and, later, North Atlantic) epistemology. The colonial difference marked the limits of thinking and theorizing, unless modern epistemology (philosophy, social sciences, natural sciences) was exported/imported to those places where thinking was impossible (because it was folklore, magic, wisdom, and the like).

(Mignolo 2002, p. 90)

Several authors assume, therefore, that the distribution of power affects the production of knowledge and try to imagine alternatives that should emerge from the "oppressed" regions of the world. In a certain sense, postcolonial studies—using different strategies and languages—impugn the underlying epistemology buried in the knowledge of the dominators of the modern world and call instead for the construction of a new type of epistemology that will function on bases radically opposed to previous ones. The critique of "Western" epistemology is relatively easy to understand and seems to be in line with a way of rethinking or updating the dependency theories of the late 1960s. On the contrary, the proposals on how to build this new epistemology offer a wide range of variants— according to its proponents—but far less clarity in their content. This is due, in part, to the notion of "coloniality" exceeding the strict framework of a geopolitical explanation (marked by the existence of dominant and colonized regions, or other similar ways of denoting it) and placing itself squarely in the terrain of questioning the various ways of exercising power, as well as the resultant epistemologies. The emergence of gender studies since the 1980s has, then, been accompanied by a questioning of the knowledge bases that have, for example, produced masculine epistemologies on a brand of gender domination.[18]

In recent years, some STS researchers in more developed countries have started to pay attention to what happens in peripheral or nonhegemonic regions. As we saw earlier, for long decades—and still today— practically all works in the social studies of the science conducted in central countries—with the obvious exception of the historical works on the expansion of European science and some studies on scientific cooperation—were focused on laboratories, institutions, groups, and individuals located in the mainstream and on hegemonic research and/or technological developments.

Following the line of postcolonial studies, several authors decided to observe the social, political, and epistemological dimensions of science and technology in developing countries and discovered STS existing in these regions. In a rather belated and extremely partial movement, some "mainstream" scholars from the STS field thus seemed to discover that, in less developed regions, there is not only production of scientific knowledge but also reflections on those processes.[19] Let us look at some of these proposals, which are not numerous. Sandra Harding (2008) proposes that postcolonial studies in STS (PCSTS) raise new questions for history, sociology, epistemology, and the philosophy of science, not just in relation to the countries of the "South" but also challenges STS studies of the "North".

Harding rightly argues that the project to break away from a value-free or science-neutral idea showed, through ethnographic studies, that "Northern science" was also a "socially situated" knowledge and that, therefore, sciences "of the North", which are heirs to the Enlightenment, were also expressions of different paths of an "ethnoscience". Harding summarizes the various contributions of postcolonial studies, in particular, what she terms four central projects. The first involves advancing "beyond inclusion" and takes as a starting point the inclusion of studies on the scientific and technological traditions of "other cultures", but going further and analyzing both the "other developments" and those contributing to the understanding of "Northern science". The second, which I advance earlier, proposes new histories, sociologies, epistemologies, and philosophies of science. She asks, for example, how have the scientific projects and traditions of the planet interacted with each other? What has each taken from the other? How did the West invent and maintain the notion of static, timeless, "traditional" societies? And what has happened to the cognitive core of the Northern sciences if it has lost its legitimacy?

Third, she raises questions about the fact that multiple scientific traditions that overlap and partially conflict with each other (as do cultures more generally) have existed in the past, exist today, and will exist in the future. Finally, questioning the relationship between scientific and technological traditions, she suggests integrating "other traditions" in the North's S&T legacies, or analyzing the asymmetric relations of scientific collaboration between those of the North and those of the South, or "to take other cultures as models for the sciences and technologies of the West" (Harding 2008, p. 150). Viewed from Latin America, it is difficult to disagree with these proposals, even though, as we will see, they are not enough to develop a more comprehensive approach to trends in global science.

Warwick Anderson has, in various texts, also discussed the question of postcolonial studies in the STS field. After reviewing the various perspectives that explicitly or implicitly deal with the issue (some of which have been discussed earlier), he concludes that actor-network theory (ANT) offers the best tools for the analysis of technoscience because it incorporates geographic dimensions but does not stop there. In fact, according to Anderson and Adams (2008), debates about what formally constitutes "science" are now far more focused on geography than on epistemological problems. They conclude that "Euro-American laboratories are no longer the most important places to study science" (2008, p. 184). According to these authors, ANT's ability to subvert the narratives was creating a disruptive substitute to the theories of modernization and dependency in science studies, particularly as it deconstructs its arguments in favor of shared cognitive norms and institutional relationships and *dissolves the fatuous distinctions between center and periphery* (Anderson 2009). Furthermore, according to him, "what made ANT especially powerful and attractive was its more general corrosive effect, as it functioned also to undermine the other challenges to modernization theory, such as dependency and world systems theory, which relied on equally linear and homogeneous master narratives" (Anderson 2009, p. 391).

Taking ANT as a starting point, Anderson and Adams believe that

> we need multi-sited histories of science which study the bounding of sites of knowledge production, the creation of value within such boundaries, the relations with other local social circumstances, and the traffic of objects and careers between these sites, and in and out of them. Such histories would help us to comprehend situatedness and mobility of scientists, and to recognize the unstable economy of "scientific" transaction. If we are especially fortunate, these histories will creatively *complicate conventional distinctions between center and periphery, modern and traditional, dominant and subordinate, civilized and primitive*, global and local.
>
> (Anderson & Adams 2008, p. 192, my emphasis)[20]

A different perspective, whose goals are nevertheless close to these views, is stressed in Caroline Wagner's book (2008) on new invisible colleges, subtitled "Science for Development". Wagner analyzes the new configurations that have arisen in recent years and identifies the emergence of scientific networks as organizers of knowledge production in the twenty-first century. Under open conditions, this emergence generates various possibilities for the exploitation of knowledge, overcoming nation-state–bound formulas typical of the previous century. She recognizes, however, that there are a series of operations in the uses of knowledge that remain in the local space and that developing countries need to establish and develop a series of institutional mechanisms to take advantage of these possibilities in order to face this problem. According to Wagner, developing countries have an advantage over developed ones: They did not create national science systems in the twentieth century, and so the bureaucracies and institutions of the twentieth century that were the hallmark of the age of scientific nationalism are not, in these contexts, a real obstacle. They are, therefore, more flexible in gearing their research systems to new scientific developments.

This optimistic, certainly tempting version presents some drawbacks: First, it decrees the "extinction of the nation-state" as a space for science.

Inversely, in parallel with the networks, there are still national policies, local applications of knowledge, and knowledge strongly linked to a given localized production and to the spaces for its effective application and use (Arvanitis 2011). On the other hand, Wagner seems to include under the heading "developing countries" only countries with a limited tradition of research and already established national systems. While she does present some very interesting cases, particularly in Africa, it is doubtful that the advantages for development are as open as she imagines. In contrast, countries such as Argentina, Mexico, Chile, Brazil, Egypt, South Africa, and others, belonging to the broader "developing" world, have "national systems" similar to those of more developed countries (often as a mimetic effect) and an important dynamic in the production of knowledge, even if this often occurs in a context of subordinate forms and with very low local use of the knowledge (Losego & Arvanitis 2008, Kreimer 2015).

To end this brief analysis of postcolonial approaches, it is worth mentioning a recent article by John Law, one of the most representative authors of ANT, whose view, in my opinion, is shrewder than the others we discussed. Law and Lin (2017) take as their starting point the fact that there is an abundance of "postcolonial" jobs in STS for those who have taken issues relating to science and technology, or technoscience, in less developed countries as their research object. This, however, seems unsatisfactory insofar as it enacts a certain fragmentation between these studies, based on current theories in the most advanced countries (like Anderson, Law uses the idea of "Euro-American") to address "cases" that are very different from those studied in hegemonic contexts. It is a good question, formulated by Latin American scholars, including myself, a few years ago in relation to the region. Thus, the proposal of Law, who analyzes certain cases in Taiwan—where everything can have another meaning, including the notions of and very division between "theory" and "practice"—is consistent with the ANT perspective he displayed for years: a notion of "postcolonial symmetry" that implies having different versions of an object, including different languages, views, traditions, and so on. According to Law, this is no easy task:

> STS is currently dominated conceptually, linguistically, corporeally, metaphysically, and institutionally by provincial Euro-American and especially English-language practices. But if we do succeed, then we will have created a plurality of intersecting STSs and sensibilities, and we will be able to say that we have undone the provincialism of STS.
> (Law & Lin 2017, p. 222)

* * *

"Postcolonial" perspectives deserve two types of comments whose assumptions are rather different: one more general, the other addressing each perspective in particular. As a general perspective, the very definition of "postcolonial" is itself partial and Eurocentric. Partial, because it implies a temporality—colony/postcolony—that is far from representing the tensions present in the less developed countries. If this definition has a meaning (in Africa or India, for example), it acquires a very different one in Latin America, where the rupture with the colonial order—at least the formal colonial order—has been going on for more than two centuries. Moreover, the idea that the less developed countries are defined by some modification of the "colonial"—be it "pre-" or "post-"—conflicts with the actual intended objective, which is to highlight these regions' subaltern position to explain more comprehensively their values, beliefs, relationships, and institutions, and, above all, to analyze the relationship with the metropolis in greater complexity. The term itself is loaded with Eurocentrism, which is intended to be questioned and at least deserves to be rethought.

The objective is a commendable one: to show "other cultures" seemingly hidden from hegemonic visions which ignored—or subjugated—gender, race, or ethnicity marks and the ideas produced in poorer regions—all visions "from below" as Harding suggested. However, it is remarkable that advocates of situated knowledge and material and symbolic determinations of science–society relationships tend to propose such an amalgam between "Northern science" (Harding) or "Euro-American" cultures (Anderson, Law), apparently viewing them as part and parcel of the same indivisible modern "package" that has been imposed—as in Basalla's thesis—and then radiated to other continents and societies. Thus, "Northern science", "West", and "Euroamerica" appear as self-refutations of their own statements.

A similar amalgam tends to be made of "Southern" countries (with the added paradox that some of them, like Mexico or North Africa, are in the "North"), as if they were part of the same (more or less homogeneous) conglomeration. Possibly the most obvious case of this operation is Wagner's (2008), who, with admirable optimism, sees opportunities where historically weaknesses and structural obstacles have been seen. Instead, the "developing", "nonhegemonic", or "peripheral" world is a multicolored mosaic of cultures, types, and levels of development, traditions, and relationships, difficult to compare other than in their most negative dimension: They do not belong to the "Euro-American" world. Furthermore, it is impossible to compare societies or regions that underwent early institutionalization of "modern science" (such as Mexico, Brazil, and Argentina in America, or the North African countries, although the latter were more directly influenced by the metropolises, or certain countries with equivalent development in Asia) to others located in the same regions (some of them analyzed by Wagner) that have a few hundred "modern" researchers at most.

It is worth noting that most of the texts mentioned by authors enrolled in postcolonial STS studies actually refer to the study of technoscience in nonhegemonic or peripheral countries, with some works by local authors—in particular, those who write in English—where research by "Euro-Americans" still predominates. These studies tend to systematically ignore all Latin American production—or other subaltern literature—produced in the last forty years. I think the reason here is very simple, and Law is the only author who says it explicitly: They cannot read Spanish and Portuguese. In all the texts I analyzed, there is not a single reference that is not written in English, and only one or two belonging to a Latin American STS researcher: What scope, then, can all these works have when they ignore one of the "postcolonial" regions with supposedly among the highest STS production? Wouldn't these authors need to be constrained in order to verify that their linguistic limitations—which Latin Americans themselves do not accept in their students—make it difficult for them to approach the issue seriously?

Close attention reveals the idea proposed by MacLeod and taken up by several authors, including Anderson and Adams, to "bury dichotomies" like "modern versus traditional" and particularly including the "center-periphery" dichotomy. They suggest, instead, focusing on concepts such as "trading zones", originally proposed by Peter Galison (1997), and the richest exchanges between actors—and they therefore intend to recover or expand ANT—which those dichotomies seem to hide. It is true, of course, that these dichotomies were frequently used to impose a mechanical view, where the dominators—"Western" science and, more generally, "Western" societies—imposed a more or less monolithic vision on their subordinates in the colonies or underdeveloped regions, thus making local cultures, conflicts, hybrids, trading zones, and other topics invisible.

I pause briefly here to remark on the geographical locations and current uses of the points of the compass. The first is the analogous use of "West" to refer to "Europe" first and foremost, and, secondarily, to "Euro-America" (North America, in other words—or more precisely, the United States and Canada). This, of course, entails an opposition with the "East", in particular the "Far East", but it leaves Latin America badly located or simply off the map, unless we apply Alain Rouquié's quite unknown witticism and view it as the "Far West".[21] The second cardinal problem is the definition—also widely circulated—of "North" and "South", which, in the separation of hemispheres, locates Northern Hemisphere countries, like Mexico, Algeria, Egypt, Libya, Cuba, Pakistan, or Iran, in the "South". Despite these obvious geographic disadvantages, both expressions are still used as analogs of "centers" and "peripheries". It is ironic, however, that some of the authors who use "North" and "South" as a reference simultaneously advocate eliminating the use of "centers" and "peripheries" as a "dichotomic distinction".

Returning to dichotomies used as rhetorical devices, it is true their oversimplification may conceal more complex "trading zones", spaces of hybridization, learning processes, and knowledge circulation than simple transfers from one context to another, among other issues. Certainly, in this sense, it seems necessary—and I agree with this proposal—to break the rigidity of these dichotomies. But we are then faced with another problem: When completely dissolving these dichotomies, and therefore all central and peripheral dimensions, the tendency is to lose sight of the fact that we are facing deep asymmetries, both in the capacity to act in different geographic or symbolic territories and, particularly, in the availability of material resources and a remarkable asymmetry of power.

Indeed, I proposed some years ago to abandon the distinction between "center and periphery" and replace it with its plurals: *centers and peripheries*. This definition, which moves away from the immediate definition of a geographic region automatically labeled "center" or "periphery", puts the emphasis on three issues that must be analyzed jointly. First is the most appropriate level of analysis to account for the sociocognitive phenomena I intend to focus on. This was the topic I discussed in the first part of this chapter. Second are the place and role of science in each society, especially including its institutions, the various actors, and the set of local and cultural relationships that frame the production processes and the use—or "nonuse"—of knowledge. Third, and most important, are the relational aspects: how, given a certain level of analysis—for example, a given scientific field—the groups of a country, institution, or region link up with other groups elsewhere in the world and what consequences we can observe.[22]

Combining these three dimensions, we can then consider the different centers and peripheries delineated within each scientific or technoscientific field, given that not all knowledge produced in "Western" or "Northern" countries is automatically "central", but can often be "peripheral" in the dynamics of a given field. For instance, a prestigious American university can be "hegemonic" in engineering or computing science but "peripheral" in anthropology or science studies.

Conversely, this scheme allows us to break with the dichotomy whereby all the knowledge produced in less socioeconomically developed countries would immediately be peripheral, as Cueto (1989) declared several decades ago, related to his concept of "scientific excellence in the periphery". It also allows us to dynamically observe the extent of the changes as they occur: We have recently shown (Levin et al. 2016) that the production of scientific articles in China, in the specific field of the nanosciences and nanotechnologies, went from being completely marginal in all countries to becoming the world's leading producer (overtaking the United States in the total number of papers) over a twenty-year period.

Taking into account the relational aspects—the way individuals and groups from different regions are linked—also enables us to incorporate in our view the areas of negotiation, mutual learning, and hybridization, while keeping a close eye on the power asymmetries in play in those relationships. Let me cite just a few asymmetries as a result of the empirical work my colleagues and I have been doing for years. In relation to the choice of agendas, theories, and methods, it is usual practice for topics to be defined by groups with greater access to resources and equipment, usually located in richer societies. Similarly, in the division of labor into collaborative projects, the tasks most frequently assigned to "peripheral" groups relate to information gathering and processing. Last but not least, in cases of asymmetric links, collaboratively produced knowledge products are generally industrialized in the richest contexts, with companies and other agents more willing or better able to productively use and industrialize scientific knowledge. We will return to these questions in Chapters 7 to 9.

Notes

1. I have borrowed the expression from Harry Collins's (1975) well-known text.
2. It is worth noting that the joint meeting organized in 2016 by the two major STS societies, the European Association for the Study of Science and Technology (EASST) and the Society for Social Studies of Science (4S), bore the title "Science and Technology by Other Means—Exploring Collectives, Spaces and Futures", and the earlier joint meeting, in 2012, had as general theme "Design and Displacement". As we can see, the notion of "knowledge" seems to disappear. Moreover, the expression "by other means" is difficult to decrypt.
3. I have borrowed from Cambrosio and Keating (1983) the idea that the discussion about the division of modern science into disciplines is something "solved" or about which it is not worth working on. The remarkable thing is that this perception, which the authors had thirty-five years ago, is now far more evident after the development and expansion that we already know in the field of social studies of science.
4. It is worth pointing out that most of the texts we are analyzing were published by the same time of Bloor's book *Knowledge and Social Imagery*, in which he suggested the well-known Strong Program and its four principles.
5. In a later text, whose origin is a conference given at the French National Agronomical Research Institute (INRA), in response to a question about how scientists can meet demands from the productive sector and S&T policies, Bourdieu (1997) claims this is a "false question" and advises them to focus on accumulating symbolic capital within their own field, preserving its autonomy and then, based upon that capital, intervening more efficiently in the public sphere. The example he offers is that of Émile Zola and the capital he earned in the literary field and his intervention in the Dreyfus Affair.
6. The notion of black box was proposed by Whitley (1972) to explain the way functionalist sociology of science dealt with knowledge, its production processes, and its methods and theories. Given an institutional framework and a set of inputs, we obtain a set of products (knowledge), but the production process is not subject to sociological analysis. Many authors took up this notion—most of the "constructivist movement"—while others discussed it, surely the best-known being Langdom Winner (1993).
7. All these concepts have been elaborated since the 1970s by authors mostly from the United Kingdom, France, Germany, and the United States. For an in-depth analysis, see Kreimer, 1999a, 1999b.
8. Some authors—Law, Harding or Anderson—observed these asymmetries some years ago and suggested certain strategies that we will discuss in the second section of this chapter.

9. Nonetheless, if, instead of considering production of articles, we consider citations received, the role of the most developed countries (the United States and the EU) remains clearly hegemonic, while the figures of "newcomers" like China and India fall dramatically.

10. For Amin, this law is expressed as follows: "a formation has never been exceeded from its center, but from its periphery. The main contradiction of the formation, which is the dominant mode that characterizes it, is not the main aspect of the contradiction. This is located in another field, namely the conflict between the center and the periphery of the system". Another influential author during those years was, of course, Immanuel Wallerstein (1974), who wrote about the world system.

11. A good example of this hagiographic historiography are the texts by Barrios Medina on Bernardo Houssay and by César Lorenzano (1994) on Leloir. There are similar works on Carlos Chagas in Brazil or Patarroyo in Colombia. However, in these countries, there are also excellent critical works about these scientists; these include Buch (2006) and Cukierman (2007)

12. For an analysis of the changes in the history of science, see Pestre (1995)

13. The abundant texts focusing on relevant issues for the study of scientific practices and their specific contexts include some extremely influential ones: Mackenzie (1981), Barnes and Edge (1982), Collins (1985); Shapin and Schaffer (1985), Pickering (1995, 1999), among many others.

14. Varsavsky devoted much of his book (*Science, Politics and Scientism*) to criticizing the role of "scientificists". According to him, "a scientificist is a researcher who has adapted to a scientific market, who has stopped worrying about the social meaning of his activity, disengaged from political problems, and fully devoted himself to his 'career', accepting the norms and values of the important international centers, as specified in a hierarchy" (Varsavsky, 1969, p. 45).

15. This expression, not explained or further developed in the article, is very striking—though not completely counterintuitive—to the Latin American reader.

16. Lafuente and Catalá question MacLeod's restriction of his analysis to the British Empire, when it might be extrapolated to other regions. They also question MacLeod's suggestion that it is not mandatory to cover certain stages: For example, national science is not a prerequisite or a necessary step in a historical development of "modern science".

17. I borrow the expression from a poem written by Uruguayan writer Mario Benedetti and set to music by Catalan singer Joan Manuel Serrat.

18. Haraway, a genuine representative of this kind of approach states, in a well-known text: "In this attempt at an epistemological and political position, I would like to sketch a picture of possible unity, a picture indebted to socialist and feminist principles of design. The frame for my sketch is set by the extent and importance of rearrangements in world-wide social relations tied to science and technology. I argue for a politics rooted in claims about fundamental changes in the nature of class, race, and gender in an emerging system of world order analogous in its novelty and scope to that created by industrial capitalism" (Haraway, 1991, p 11).

19. A few years earlier, there was already a certain (sporadic) interest from European or American STS researchers in studies in developing countries. See, for example Moravcsik's (1985) and Hill's (1986) response and, later on, Drori (1993). It is noteworthy that while Moravcsik aims to create an agenda for STS studies, the other two scholars tend instead to formulate political recommendations.

20. Without mentioning a "postcolonial" approach, other texts also mention the abandonment of "center–periphery" models. MacLeod (2000) suggested this

and proposed instead to study the traffic of ideas and institutions, recognizing reciprocities and using perspectives "colored by the complexity of contact". Secord (2004, p. 669) proposed that the history of science be moved beyond ANT to "a more complete understanding, often nourished by anthropological perspectives, and [should] replace the divisions of center and periphery with new patterns of mutual interdependence".

21. Rouquié (1998, p. 16) very elegantly points out that "Latin America appeared for a long time as the third world of the West or as the West of the third world. Ambiguous place in which the colonized is identified with the colonizer".

22. *Noblesse oblige*, the only author among the texts we discuss earlier to emphasize the role of international collaborations and their consequences is Harding. Even though Wagner also analyzes international networks, she sees these links more as an opportunity than as a risk for developing countries.

3 Social and Scientific Problems—A View from the History of Science
Chagas Disease as a Model (Part 1)

Introduction: The Social Utility of Scientific Knowledge

In this chapter, I discuss one of the central topics in "science–society" relations in Latin America, which can ultimately be encapsulated in the following rather brutal question: What is the region's scientific production for? As I mentioned in the previous chapters, one of the historical issues—perhaps *the* historical issue—of Latin American science is its extremely low use—social, economic, or even for public regulations[1]—of locally produced knowledge.[2]

Most actors in Latin America (public authorities, academic communities, international organizations, and civil society organizations) express the belief—in a certain explicit discourse, at least—in the development and application of scientific knowledge being able to help to overcome—or, at any rate, alleviate—various social problems. This is expressed in numerous explicit scientific policies that place heavy emphasis on the notion of the *social relevance* of knowledge. Almost all plans from science and technology planning and management agencies have demonstrated this for several decades now.

I intend in this chapter to analyze these relationships by highlighting the different aspects involved, such as social ways of public theming and articulation of social problems, strategies for "mobilization" of scientific knowledge as a strategy for addressing these problems, and the role of scientific knowledge itself in the definition of public discourse and policies. These issues are necessarily accompanied by others: the local history of research traditions in different scientific fields, the tensions between social uses of knowledge, and the relationships with the international scientific mainstream, tensions that are historically present in peripheral contexts.

To illustrate these questions I will consider a very significant case: the double emergence (or coproduction) of Chagas disease as a scientific topic and as a public problem from a historical point of view.

I am particularly interested in looking into this issue in Latin America because the dynamics of these processes in peripheral contexts usually present specific features that are, at the same time, of special importance for social research and highly sensitive for societies in the region. As I pointed out in the previous chapters, one characteristic of science in peripheral countries is the scant effective utility (appropriation by other actors) of scientific knowledge as compared to what happens in core countries, where locally produced knowledge, for the most part, generates innovation, improved productivity, well-being of the population, global competitiveness, or environmental improvements.

Insofar as I believe that "social problems" do not emerge independently of who (which actors) thematizes them as such, I wish to highlight the way Chagas disease has been historically constructed as a "public problem", the stances of different actors around the disease, the actions undertaken by them, and, especially, the strategies that stimulate oriented scientific knowledge production as a useful (and legitimate) tool to approach it.

* * *

Understanding the processes whereby scientific knowledge acquires a (social) utility that goes beyond the legitimacy granted by the scientific community is a central issue for the social studies of science and has been analyzed from a variety of perspectives. One very common view involves the image of science as an enterprise that, on certain occasions, is "oriented" to social demands that present themselves as pending issues. This will of scientists to "connect" their practices with other areas of society has produced numerous studies showing the different dimensions present in these processes: economic interest, political commitment, response to funding opportunities, reproduction mechanisms in the scientific community's research traditions, or deeper transformations of knowledge production processes.

Notwithstanding the differences between these different perspectives, a common element can be observed: All of them show a central concern about how knowledge is produced and the way it is (or should be) affected by the construction of its social utility. Thus, by concentrating exclusively on knowledge production, these studies have partially left aside the processes of formulating the demands, that is, the way certain issues acquire the status of a "social problem". Therefore, it is possible—and legitimate—to deploy a variety of strategies to solve them. In other words, social problems are treated as "data", the nature of which it is not necessary to approach from the sociology of knowledge. Moreover, only a few authors consider the process whereby social needs may eventually become "demands" for knowledge or issues that can be addressed through scientific research.

On the contrary, my work has adopted a perspective according to which *the production of scientific knowledge participates in defining and imposing certain topics on the social agenda.* Accordingly, if we are interested in understanding how scientific knowledge becomes useful to society, it is not enough to observe research practices developed *once the topic has been installed.* In fact, as Shapin and Schaffer (1985) point out, "solutions to the problem of knowledge are embedded within practical solutions to the problem of social order, and that different practical solutions to the problem of social order encapsulate contrasting practical solutions to the problem of knowledge".

I am thus oriented by a theoretical assumption that the emergence of the problem, the definition of the practices proposed in each period to solve it, and the decision to allocate resources for such practices to be carried out are all the result of interactions between different social actors operating within certain institutional frameworks that contain them, while, at the same time, "shaping" their actions and interests. Accordingly, insofar as it acquires visibility and becomes a public object, the "problem" translates into a series of stances, the enlistment of other actors, the creation of institutional arrangements to address it, and the practices associated with such arrangements, which simultaneously condition the type of knowledge produced and its possible use.

This theoretical position is, in a way, close to Pierre Bourdieu's idea: "The public expression of social needs is nothing but a euphemism to conceal the private (economic) interests of companies or big industrial groups" (Bourdieu 2001). Although this statement seems exaggerated (there are, in fact, broader interests than private economic ones), it is important to emphasize the idea that "social needs" do operate as a euphemism mobilized by other actors to impose their own point of view on the object in question.[3]

This approach leads us to believe that the very idea of "social needs" is problematic: It is crucial to determine how actors vie for the always difficult task of establishing "who has a legitimate right to speak 'on behalf of the poor'" (Bourdieu 1997). Precisely, the most disadvantaged social actors in a peripheral context are also those who have greatest difficulty "expressing" their needs in terms of real or potential scientific knowledge. In fact, the population affected by Chagas disease (considered a "disease of poverty") has *not been established as a relevant social group in the fight against the disease, their needs always being "translated" by other social actors.*[4] The actors directly affected by the disease are, therefore, those least capable of expressing themselves in the public arena, while laboratories (public and private), doctors, health authorities, and, of course, scientists operate in public disputes according to their own interests and with weapons legitimized as such.

Three broad expressions of social needs can be seen in the public space:

1. Most of the time, the "voiceless" are represented by the state, which determines their "legitimate" social needs and, among these, the ones

capable of being addressed by scientific research. However, this is not a matter of the state as an "idealization of the common good", but a cross between a bureaucracy composed of public employees and scientists, who, as advisors of officials, become "spokespeople" either for the scientific community itself (or a portion of it) or for the international networks they are part of.

2. On the other hand, it is often the scientists themselves who establish— either in purely rhetorical or real ways—the "possible applications" of their research, which should act as legitimation mechanisms for funding agencies or other actors. In this type of justification, there is always an explicit or implicit identification (construction) of "social needs", which should legitimize their research.

3. Third—and this becomes increasingly important in peripheral countries—a group of international funding agencies establishes a "list of social priorities" as a condition to grant credits for scientific research. A list of topics needing to be tackled by research is then derived from this.

Consequently, to fully understand the problem, we must keep in mind the social dynamics—where scientists are clearly not the focus of disputes— and the strategies of other actors, including "laboratory" practices, who mobilize heterogeneous resources (both material and symbolic) and circumstantial allies.

Different Actors, Different Kinds of Intervention

To plot the representations of different actors and, therefore, of different logics and ways of "acting" on the problem, we assume the principle of "interpretative flexibility", which, during our inquiry, forces us to take into account the points of view of different actors and to vary our own vantage point in order to understand the various logics involved. The emergence of the "problem" is not a univocal, predetermined process; rather, the different interventions shape it as such.[5]

These interventions are based, in turn, on various representations of the problem simultaneously originating in the social dimensions and epistemic aspects of the knowledge object. The social and cognitive aspects of the process studied are therefore inseparable.

Applying the concept of "interpretative flexibility", which has implications for both theoretical and, especially, methodological issues, I have identified different groups relevant in the construction of Chagas disease as both a social and epistemic problem. The first stances consistently construct a first "artifact": the link between a previously unknown causative agent (pathogen)—*Trypanosoma cruzi* (hereafter, *T. cruzi*)—with clinical symptoms attributed to different ailments (goiter, cretinism). But when the scene shifts from Brazil to Argentina, it is other social actors who, *from another actor's perspective*, construct the second artifact, formed by the link between *T. cruzi* and a new disease: Chagas.

To summarize, in this chapter, I argue that:

1. It is not possible to consider a single social situation as intrinsically problematic. On the contrary, it has to be related to the role of social actors that construct it in situations that are historically contingent.
2. In contrast to the normative naturalization of the role of knowledge in solving social problems, I am interested in the reciprocal investigation of the role of scientific knowledge in constructing social problems, perhaps as a mark both of a new modernity based on the strength of science and the role of scientific knowledge in reorganizing societies.
3. Knowledge is itself the product of social constructions. The interventions and the use of knowledge in other fields of symbolic and material production mold its social role and its cognitive content as well.

This allows me to analyze, using the same matrix, the production and use of scientific knowledge, and, therefore, I am able to map the "trajectories" of knowledge. In fact, if we acknowledge that the use of knowledge is the result of incorporating a certain knowledge into the usual practices of a group of actors, the use of scientific knowledge would be the result of a *sociocognitive articulation* capable of appropriating this knowledge and reconfiguring it according to its own interpretations. This arrangement comprises a set of actors, with their respective institutions providing them with material support for their activities and the content of the knowledge in question.

For example, the existence of scientific research on Chagas disease is the result of the scientific community (with the consequent mobilization of political resources) incorporating "knowledge" that, at the level of the social representations, postulated Chagas disease as a social problem. Far from offering a merely passive response to a given problem, the scientific community played a key role in these processes, *shaping* the problem, rather than passively "receiving" any demands that might come from the civil society.

In the perspective I adopt here, the more strictly cognitive dimensions, such as the nature of the disease and its symptoms, the physiology of the parasite and the ways it infects humans, the conditions under which the vector (the *vinchuca*), etc., operate are inseparable from the ways in which doctors, scientists, public authorities, and other actors have defined and acted on the problem.

In analyzing this issue, we have to remember that the social problem and the populations affected (or at risk) of Chagas disease have no independent existence from the constellation of actors and institutions that represent and define it and operate on and negotiate its development.

The Historical Trajectory of Chagas Disease as a Social and Knowledge Problem

Here, I lay out the historical evolution of Chagas disease as a problem of knowledge and its establishment as a social problem. Following this story, it is possible to observe how, in the complex process involved in the production (and acceptance) of an item of scientific knowledge (in this case, the existence, scope, and characteristics of Chagas disease), heterogeneous actors intervene (mainly scientists, government officials, and international agencies) and how displacements occur both in terms of disciplinary fields (from protozoology to molecular biology) and of the scope, dimensions, and characteristics acquired by the social problem.

From Ranchos to Laboratories: The Discovery and Rediscovery of Chagas Disease

The process of recognizing Chagas disease as a regional epidemic is atypical in the history of medicine. Unlike other epidemics, like yellow fever or tuberculosis (Barnes 1995; Coleman 1982; Gilman 1988), where the effects of the disease on people—and, as Löwy (2001) points out, on the economy—preceded the interest of researchers looking for scientific explanations to allow its control, Chagas disease emerges as a result of the practice of "normal science" within the protozoology of the time (Benchimol & Teixeira 1994).

In 1909, Carlos Chagas, of the important Oswaldo Cruz Institute in Rio de Janeiro (known as "Manguinhos Institute"), Brazil, was sent on a mission to control malaria in a rural area in the center of the country. There, alerted to the existence of a blood-sucking insect (the *Triatoma infestans*, familiarly called *vinchuca* in Argentina, *barbeiro* in Brazil, and *pito* in Colombia and Venezuela), Chagas discovered a hitherto unknown trypanosome inside it (later baptized "*Trypanosoma cruzi*" in honor of his teacher and director at the Institute, Oswaldo Cruz). He soon incorporated this in the model of tropical diseases: a causal agent, a transmitting vector, a pathology.

Thus, Chagas made a detailed description of the acute form of the disease, namely, those cases where the parasite infection causes a reaction in the body (fever, enlargement of the spleen and liver, swelling of the face) that can be fatal. He achieved this by recording cases of sick people and confirming these data with laboratory experiments which reproduced the processes in animals inoculated with *T. cruzi*. He also associated the characteristics of the chronic stage of the disease (when infected people do not manifest the symptoms of the acute stage or these symptoms have disappeared) with the numerous cases of goiter (hypertrophy of the thyroid gland causing a thickening of the neck) and cretinism seen in the area, identifying the pathology as *parasitic thyroiditis*. (Delaporte 1999).

The discovery was an event in Brazilian society. On the one hand, the scientific recognition brought the Manguinhos Institute to the fore in the field of protozoology, dominated at the time by the German school, with which Brazilian scientists were in close contact, and brought Chagas major scientific and institutional recognition.[6] On the other hand, insofar as Chagas disease was associated with goiter and cretinism, endemic in much of Brazil, it signified the recognition of a major public health problem.

However, the initial enthusiasm with which the discovery was met soon tailed off. A working paper submitted by Rudolph Kraus, director of the Bacteriological Institute of Buenos Aires, was fundamental: It showed that, in certain regions of northern Argentina, despite the enduring presence of the parasite, goiter was not observed, challenging Chagas's identification of the associations between the infection and *T. cruzi* and symptoms of hyperthyroidism. The inability to empirically confirm the relations established by Chagas (which resulted in severe clashes in the medical field and Brazilian public health regarding the existence and importance of the epidemic) led to a gradual loss of interest in the disease in ensuing decades at the levels of both health policy and science (Coutinho 1999; Kropf, Azevedo, & Ferreira 2003).

The history of the disease then shifted to a different geographical space, different actors, and different disciplinary fields. In 1929, Salvador Mazza, an Argentine doctor specializing in bacteriology, was appointed director of the Mission for the Study the Argentine Regional Pathology (MEPRA) in Jujuy Province, after three years under José Arce, head of the Surgical Clinic Institute of the National Clinics Hospital, belonging to the University of Buenos Aires (UBA). Its attachment to the UBA aside, the MEPRA had, from its inception, the support of the governor of Jujuy Province and the local ruling classes, who donated a house that served as the mission's headquarters. Its main objective was to determine the extent of northern Argentina's different pathologies, though it would soon concentrate most of its efforts on Chagas. Over the seventeen years Mazza was in charge of the mission, the research allowed him to chart the people and animals infected with *T. cruzi* in the area, the infestation of *vinchucas* in households, and the characteristics of the chronic form of the disease—a crucial point in its recognition. The recognition that certain people developed an ocular edema on infection (known as "Romaña syndrome" in honor of the doctor who first identified it) was crucial. It proved to be the key to the rapid diagnosis and clinical description of the disease in its acute condition and allowed numerous cases to be confirmed, putting an end to doubts about the extent of the disease (Delaporte 1999).

Thus, the recognition of cases of the disease saw exponential growth between 1935 and 1940 in both Argentina and Brazil, where a group from the Oswaldo Cruz Institute based in Lassance, to which Carlos's son Evandro Chagas belonged, had continued research into the topic.[7] The extent of Chagas disease was soon recognized, as it went from being termed *parasitic thyroiditis* (as Chagas had proposed) to *American trypanosomiasis*. The main research then shifted to the clinical field—particularly to cardiology—making it possible, over the 1940s and 1950s, to determine that the main characteristics of its pathology were lesions to the heart and digestive tract.

From Laboratory to Desktop: The Institutionalization of Chagas Disease as a Social Problem

Starting in the late 1940s, the disease leapt the barriers of the scientific field and was established as a social problem. Around it developed a significant institutional apparatus, ranging from programs to combat the vector to research and technology institutes charged with developing and refining existing diagnostic methods.

One of the first manifestations of the growing importance of Chagas disease as a research object was the creation of the Institute for Regional Medicine in 1942, attached to the National University of Tucumán (and later to the National University of the Northeast), which had "missions" in the provinces of Chaco (1945) and Jujuy (1947). From its foundation, the institute was headed by Cecilio Romaña, who had worked closely with Salvador Mazza in the MEPRA (from whom he later distanced himself) and with researchers of the Oswaldo Cruz Institute in Brazil, a relationship that would mark his scientific work. In this period, when the MEPRA moved to Buenos Aires and subsequently closed, different research was conducted at the Tucumán Institute, which allowed a more precise characterization of clinical symptoms, diagnostic methods, and the epidemiological extent of the disease. From there, the first trials with hexachlorocyclohexane were conducted, an insecticide capable of killing *vinchucas*, whose effectiveness had recently been proven in Brazil.

The knowledge produced in these last two aspects—the epidemiological inquiry and household disinfection trials—were decisive in the formation of Chagas disease as a social problem, a process in which Romaña played an important role. In fact, the "First Pan-American Meeting on Chagas Disease", organized by Romaña in Tucumán in 1949, was the first to discuss the findings of research into the viability of controlling the presence of the *vinchuca* in the region's households. A special guest at this meeting was Ramón Carrillo, President Juan Domingo Perón's first Minister of Social Assistance and Public Health. For various reasons, Carrillo was particularly sensitive to the topic, which did not enjoy much recognition at the time.[8]

One of the main consequences of the Tucumán meeting was the creation, under the Ministry of Health, of the Steering Committee for Chagas

Disease Research and Prophylaxis (Zabala 2010) in 1950, the coordination of which was entrusted to Cecilio Romaña. With the exception of the university research institutes mentioned earlier, this was the first institutional manifestation of Chagas disease as a public health problem and started a process whereby the disease acquired the status of a "social problem".

On the steering committee's recommendation, the National Service for Prophylaxis and Resistance against Chagas Disease (SNPLECH) was set up in 1951, also under Romaña, and made its headquarters in Buenos Aires in 1953. In the framework of these institutions, measures were undertaken to control the vector (household fumigations) as well as research which led to exponential growth in the recognition of sick people. The epidemiological surveys carried out by Mauricio Rosenbaum and José Cerisola from the early 1950s were key in this regard. These surveys combined diagnoses of infection with electrocardiographic tests, thus confirming the epidemiological relationship between infection with the trypanosome and cardiomyopathy. It implied an exponential leap in the recognition of the disease through the detection of chronic cases (in which the parasite's presence in the blood is rare, and the disease's symptoms are located in the heart). Indeed, until 1946, the year of the MEPRA's move to Buenos Aires, Mazza and his collaborators had recorded 1,400 cases of Chagas disease 1,100 of which showed the parasite in the blood (Sierra Iglesias 1990). With Rosenbaum and Cerisola's surveys, these numbers grew and grew, with the number of infected people being calculated in the hundreds of thousands (Rosenbaum & Alvarez 1953; Rosenbaum & Cerisola 1957, 1958)

In 1957, Cerisola was appointed director of the first Chagas laboratory attached to the SNPLECH, where he began to conduct research on the trypanosome and various species of triatomines. After a series of institutional changes, this laboratory became the basis of today's Mario Fatala Chaben National Institute of Parasitology,[9] the main nonuniversity institution devoted to research and diagnosis of the disease.

In the same period starting in 1958, actions to combat the vector that had started out as trials a few years earlier were undertaken systematically. They were consolidated with the creation of the National Program on Chagas Disease (PNCh) in 1961, still operating to this day. Notwithstanding this continuity over time, the actions of the PNCh were not consistently systematic, due largely to intermittent funding, and often showed a shortfall in resources to meet sustained prevention (due, among other factors, to an aging staff, a lack of vehicles and fuel consignments, and a shortage of insecticides).

At any rate, this period was marked by institutional development around the disease which, in turn, entailed a recognition of the importance of Chagas in terms of public health.

Back to the Labs: Molecular Biology Comes Onto the Scene[10]

In parallel with institutional recognition, there was growing interest in Chagas from scientific researchers, as shown by the creation in 1970 of the UBA's Commission for Clinical Research into Chagas, which coordinated research in biochemistry, microbiology, and medical clinic. However, unlike previous stages, where most research was conducted in the field of medicine, the center of gravity of scientific research shifted, from the 1970s, to the field of biology. This research comprised, on the one hand, studies of the vectors in close connection with the control strategies implemented from the 1960s; on the other hand, there was the development of a whole tradition of research in molecular biology, which, from the 1980s on, took the parasite as its object of study and whose ultimate goal was to develop a vaccine. The hegemony of molecular biology groups in the distribution of the various different disciplines that took some aspect of Chagas disease as an object of study has survived into the present, as can be seen in Figure 3.1.

Coutinho notes that, in the case of Brazil, "the success of molecular parasitology groups was linked to their ability to attract resources, their institutional characteristics, and their international connections" (Coutinho 1999). Something similar could be said of Argentina, as we have shown (Kreimer & Lugones 2003; and especially Kreimer 2010d), when looking at the beginnings of molecular biology. At the institutional level, this growth was led by the groups embedded in the biomedical tradition passed down by the Nobel Prize winners Houssay (in 1947) and Leloir (in 1970), who achieved a strong (and, in that sense, "successful") connection with the international community, as expressed in researchers

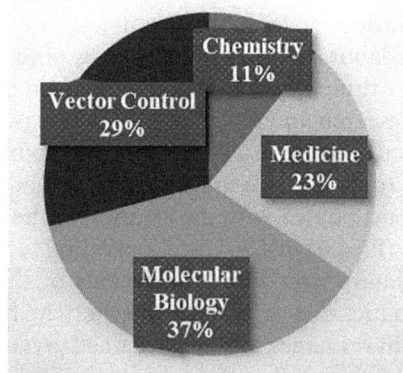

Figure 3.1 Percentage of Research Groups Devoted to Chagas Disease by Research Area in Argentina in 2004

Source: Author's own, based on data from scientific promotion agencies and universities in Argentina

trained overseas and international journal publications.[11] On the material plane, prestige brought researchers sources of stable resources, from national and international agencies alike. At the national level, there was the National Program of Research into Endemic Diseases, launched in 1974 and serving as a stimulus for academic research in biochemistry and molecular biology on Chagas disease.

Keeping in mind the formation of local research traditions, it is noteworthy that, from the very beginning of the (clearly innovative) development of molecular biology in Argentina, the study of *T. cruzi* formed part of that new field. In fact, virtually all of the most prestigious groups devoted to studying the parasite from the perspective and techniques of molecular biology were trained at the then Campomar Foundation, today the Leloir Institute. The pioneers in this regard were researchers from the first generation of Leloir's disciples, such as Héctor Torres, who, at the start of the 1980s, left the Campomar to found another emblematic institution, Campomar's "daughter", the Institute for Research on Genetic Engineering and Molecular Biology (INGEBI CONICET-UBA). The groups formed would become absolutely central in the study of various aspects of Chagas, in particular, various mechanisms present in the parasite that—and this would become central to the groups' strategies— were then taken as an important biological model. Thus, Torres would shift from certain problems more typical of biochemistry to the study of signal transduction in *T. cruzi*.

Armando Parodi is another of the emblematic researchers in these areas. As early as the 1980s, a few years after Torres—again at the Campomar Foundation—he devoted himself to the study of the protein glycosylation in *T. cruzi*. His early works on this subject were associated with Juan J. Cazzulo, another researcher who also focused on the construction of similar objects. Along with Parodi, Carlos Frasch, and Rodolfo Ugalde (all of them trained at the institute set up by Leloir), Cazzulo set up another Campomar "daughter" institution: the Institute for Biotechnological Research (IIB-INTECH), at the National University of General San Martín, in the early 1990s. Among other issues, they studied the trans-sialidase enzyme in T. *cruzi*—and the regulation of gene expression in the same parasite.[12]

In parallel, several researchers trained at the Institute for Research on Genetic Engineering and Molecular Biology (INGEBI CONICET-UBA) worked on aspects of Chagas disease and particularly on the study of the trypanosome. The man who has devoted himself to them most consistently has been Mariano Levin, who for years has been working on the structure and function of the antigens present in *T. cruzi*. Actually, in this institution, the study of *T. cruzi* as a research object has exercised a number of researchers besides Torres and Levin: people like Téllez de Iñón or Mirtha Flawiá, among others, so that, as in the IIB, it forms a historical and traditional line deeply embedded in the institution's historical structure.

Finally, at the international level, the creation of the World Health Organization's Special Program for Research and Training in Tropical Diseases (TDR) in 1975 was particularly important. It meant a fundamental support for the consolidation of molecular biology research into Chagas disease. This was accentuated in 1994 with the launch by the TDR of the "*Trypanosoma Cruzi* Genome" project. This program, involving twenty laboratories working as a network, was aimed at the complete sequencing of the parasite's genome, thus ascertaining the structure of the molecules involved in the infection. Argentine laboratories played an important part in it: Of the twenty laboratories, three were Argentine (the IIB-UNSAM, directed by Carlos Frasch; the INGEBI, directed by Mariano Levin; and the Fatala Chaben Institute). The other laboratories break down as follows: nine in Brazil, one in Venezuela, one in the United States, and seven in Europe (Germany, Spain, France, the United Kingdom, and Sweden). Using this information, the intention was to identify "possible" targets in the parasite that will eventually lead to the development of therapies and/or forms of prevention (such as a vaccine).

These expectations, however, are more of a rhetorical device than true applications to solve the problem, as the research has been carried out exclusively in the field of basic research and—to date, at least—there have been no research groups with the skills needed to develop such medicines.[13] Chagas's relevance as a problem, then, is confined to the canons of international science.

Ecology and Other Subfields

In terms of the production of knowledge linked to combating the vector, there are several research groups, three of them in the University of Buenos Aires located in the departments of Ecology, Genetics, and Evolution, Biodiversity, and Experimental Biology; one in the National University of Córdoba; and one in the Armed Forces Scientific and Technical Research Institute (CITEFA).[14] The first group, directed by Ricardo Gürtler, is "Population biology of insects, pathogens, and vertebrates". Specializing in bioecology, its main analytical focus has, since the mid-1980s, been the study of the ecology of vectors of Chagas disease in northern Argentina, specifically in rural areas of Santiago del Estero. Since its inception, they have incorporated lines of research that bring them close to other researchers in areas including the application of molecular markers to understand the "spatial and temporal structure of the population of both parasite and vector", as well as the use of satellite technology to study the transmission process of infestation, among others.[15] The second of the groups, "Ecology of reservoirs and parasite vectors", is directed by María Cristina Wisnivesky and focuses its research on the ecology of triatomines, although in recent years they have been adjusting their research agendas and moving away from *T. cruzi* as a problem. The third group, at the University of Buenos Aires, is the "Laboratory of Insect Physiology" (Biodiversity and Experimental Biology), directed by Claudio Lazzari. Their research is both the physiology and behavior of *vinchucas* and the insect–parasite relationship.

In Córdoba, one of the provinces affected by the disease, a group from the Faculty of Chemical Sciences does research into the biochemistry of the *vinchuca*, following lines of work related to processes in which the insect has to consume a great deal of energy and others related to the biochemistry of the parasite–host interaction. In the first case, they have studied flight mechanisms, identifying the molecules involved in the transfer of energy to flight muscles.

Finally, a research group is based in the CITEFA within the Center for Pest and Insecticide Research, notable for the development of a fumigator pot, a simply applied insecticide.

It is important to note that, unlike the groups focusing on *T. cruzi*, the centers involved in the research on the vector (the *vinchuca*) have an explicit strategy of crossing the borders of laboratories and of interacting more actively with other actors. There are several reasons for this: On the one hand, embedding within the disciplinary field of ecology, whose borders have been more comprehensively established than in other disciplinary fields. On the other hand, it can be explained by the eminently practical nature of most of this research, where field trials form the main focus of the strategy.

Structure of Scientific Knowledge Production in Chagas Disease

In this section, I briefly analyze several dimensions to give us a more comprehensive picture of Chagas research: first, the geographic distribution of research, correlated to the zones of higher prevalence of sick people; second, an overview of the scientific production, taking papers published in international scientific journals as a main source (through citations provided on three databases). I then look at the thematic distribution of scientific production. Finally, I focus briefly on current research dynamics in Chagas disease.

Concentrated Geographic Distribution

To enrich our understanding of Chagas disease both as a research topic and as a social problem, it is worth observing the geographic distribution of research groups working on Chagas disease across Argentina. We have to correlate two variables: the location of research groups and the distribution of endemic areas, the latter being concentrated in the (poor) northern provinces. Let us look at the result of crossing these two dimensions (Figure 3.2). The dots mark the existence of Chagas research

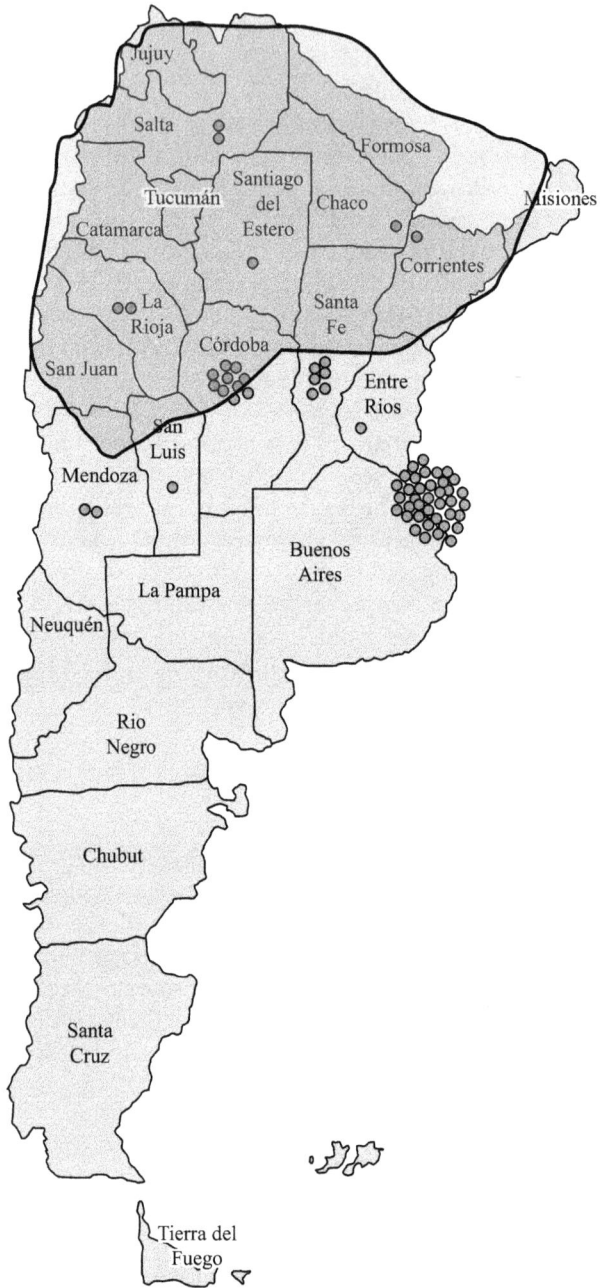

Figure 3.2 Distribution of Chagas Research Groups and Endemic Areas in Argentina

Source: Author's own

groups, the shaded area, the areas of greatest preeminence of sufferers and vector transmission.

It is easy to see that the vast majority of groups are located in large urban centers (Buenos Aires, Rosario, Córdoba), more or less away from endemic areas. The map speaks for itself so clearly that it makes any further comment redundant.

Important Scientific Production

The quantitative importance of research not only refers to the number of groups working on the subject but also to the production of articles published in international journals. In other words, there are many groups working on the subject, and they are, relatively speaking, highly productive.

Publications by Argentine researchers—or those based in Argentine institutions—were surveyed, using three citation indexing databases, over a ten-year period (1995–2005) in the most important international journals and according to a set of keywords (Chagas disease, *Triatoma infestans*, *vinchuca*, *T. cruzi*, Chagas cardiomyopathy). The result is a highly significant number of articles cited in those databases (Table 3.1).

Thematic Distribution of Production

Scientific research related to Chagas disease has covered multiple aspects linked to the disease's development, including its conditions of emergence, transmission, diagnosis, treatment, and prevention.

The map of scientific production on Chagas shows us different cognitive concerns/research objects, disciplinary approaches, techniques used, and institutional spaces for knowledge production.

We have established a classification for Chagas research according to the different cognitive reference objects: sick people, the parasite (the causal agent), the vector (the *vinchuca/Triatoma infestans*), spatial distribution, and so on.

Table 3.1 Publications by Argentine Scientists in 1995–2005, by Database

Database	Number of papers
Science Citation Index	830
Medline	650
Biological Abstracts	170
Total	1,650*

* N.B. The total should not be understood as a simple sum, since some of the citations in the three databases may overlap. The total only represents the number of quotations corresponding to the period.

Taking publications in the Science Citation Index as a reference, we can observe their thematic distribution (Table 3.2).

As we can see in Figure 3.3, half of the surveyed publications are about the parasite (*T. cruzi*). There are two reasons for this: on the one hand, the heavy concentration of research in molecular biology and biochemistry that takes *T. cruzi* as a research object; on the other, the effects of the WHO–TDR–funded genome sequencing program for the parasite.

On the other hand, it is necessary to emphasize that many researchers in molecular biology have taken the parasite as a *biological model*, that is, to say, for the study of certain particular biological mechanisms

Table 3.2 Thematic Distribution of Publications by Argentine Scientists in 1995–2005 as Indexed in the SCI

Object of study	Number of papers	Percentage (%)
Parasite	415	50
Patients	191	23
Vector	183	22
Epidemiology	33	4
Others	8	1
Total	830	100

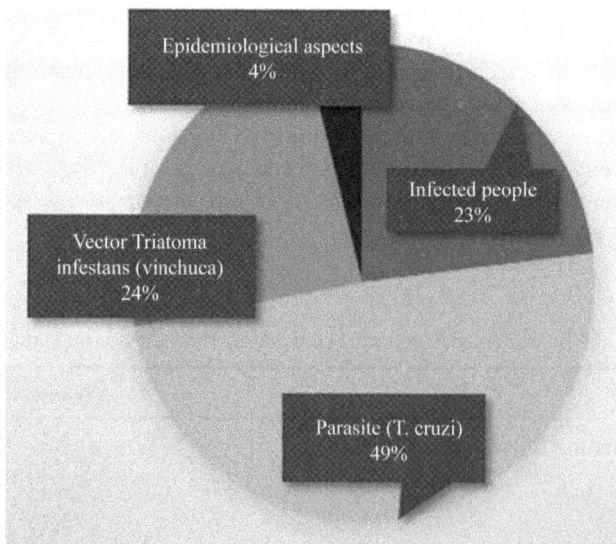

Figure 3.3 Thematic Distribution of Papers Published on Chagas disease (1995–2005)

to be found in this type of organism. The relationship of this important research with Chagas disease is therefore problematic (I come back to this in my conclusions).

Approximately one-quarter of production is oriented to sick people. Within this thematic orientation, the study of basic aspects of the disease predominates, while clinical research occupies a very insignificant part. This is consistent with the distribution of research according to institutional spaces, with only a small portion being conducted in hospitals and other health centers.

Dynamic Aspects of Chagas Disease Knowledge Production

As we have shown, research into Chagas disease displays significant thematic and institutional heterogeneity. Therefore, to understand how this research relates to the diagnosis, treatment, and prevention of the disease, it is necessary to analyze the dynamics of the relations between researchers and other actors who carry out these practices.

Strong Predominance of Basic Research

There is a strong predominance of basic research in all fields of knowledge involved. This research is, in general, directed at learning about different features of the parasite (its reproduction mechanisms, elements involved in the interaction with its host, genetic classification of various strains), the vectors (eating habits, reproduction dynamics, genetic characteristics, morphology), and the interaction between the parasite and mammals (immune response of those infected, organs affected).

The chief product of this research is scientific papers, most in international journals. In turn, the main (almost exclusive) sphere of circulation and diffusion of this research is the field of academic research through scientific publications. This implies a limit on the capacity to disseminate this knowledge, which acquires an endogamic character insofar as understanding it requires a high level of specialization only possessed by highly specialized scientific researchers.

Good Linkage Between Scientific Research Groups and International Networks, But Limited Between Local Groups

Most of the research groups maintain good links with the international community, especially the United States and Europe: A majority of researchers participate in international networks and joint projects funded by international agencies (National Institutes of Health [NIH], World Health Organization [WHO], Howard Hughes, the European Union, and so on). In general, the origin of these links goes back to the training overseas of the group's director and is reinforced by the exchange of new generations of researchers. These relationships translate into joint publications between national and international groups, often in areas of interest for laboratories in central countries.

There is also another type of linkage arising from thematic affinities in research. In this case, cooperation is usually the result of researchers' initiatives when faced with certain specific needs in their research (the provision of a piece of equipment or certain research inputs). This type of relationship is more frequent in the countries of our region (mainly Brazil) and is of variable duration. Sometimes, it only involves a specific cooperation; at other times, more permanent relationships are established.

At the same time, relations between research groups in Argentina are thin on the ground, either between research groups working in similar areas or between those belonging to different research fields. Several reasons can explain this phenomenon: in some cases, the struggles for priority among the groups sharing the research topics; in others, the lack of connection is usually attributed to opposing perspectives within the research. And finally, in some cases, lack of communication is the result of personal conflicts between researchers, leading to their respective research groups developing parallel pathways. As a result, relations among groups tend to be couched in terms of competition or, at best, indifference.

Lots of Knowledge Production, But No Useful Knowledge or Applications: A Mark of Peripheral Science?

However, despite the sturdy development of research into various aspects of *T. cruzi* and the *vinchuca*, particularly over the last two decades, the effective utility of knowledge to solve the social problem has been low. Strikingly, there has been no significant contribution either in elements for the production of vaccines or in the replacement of two traditional medicines to treat the disease (benznidazole and nifurtimox), nor, finally, any likelihood of a new drug being introduced into the market.

There is no single cause to explain this phenomenon. Undoubtedly, there are a number of contingent factors: for example, the type of construction the different perspectives have created around the disease; the economic, institutional, and political conditions; or the different logics of action in play (researchers committed to academic careers, the low recognition of professional doctors devoted to the disease, the lack of long-term institutional policies). But it is also necessary to add a structural element, which is the development of science in peripheral contexts, a process I have conceptualized as "subordinate integration". This is apparent in the case of scientists who have done research into Chagas, among whom there predominates a logic of international collaboration and integration with research laboratories in the mainstream of international science, rather than a true orientation toward treatments for the disease. This type of integration has resulted in local research agendas being more closely aligned with problems and objects that bring greater visibility and international scientific legitimation than with the development of knowledge products designed to respond to local problems.

"Purification" Processes: The Utility of Knowledge in a Peripheral Context

If we look at the sequence of appearance and the dynamics of the actors involved in the process described, three aspects are evident. The first is that the "modern" perspective implied by molecular biology has redefined the terms of the problem, setting in motion a process which I term "purification": sick people, their conditions, environment, housing, and so forth are blurred, first by physicians who take their blood, and then by biologists who extract parasites from it, sequence their DNA, and study the regulation of gene expression, for example, or other relevant biological processes. These shifts of purification can be schematized as shown in Figure 3.4.

By the time we reach the last dark square inside the oval for the international scientific community working on specific problems, all the conditions present in the natural and social context have disappeared and their reconstruction makes no sense within this new configuration of actors. Except, of course, when it comes to justifying the relevance of this knowledge: In that case, a symbolic relationship is established between the knowledge and the production of "solutions" for the disease. But "disease" is not the same as "sufferers", and therein lies one of the keys to this operation of symbolic reconstruction.

As we can see, at the end of the road, if a drug or new treatment will—someday—be produced, it is more than possible it will be developed by transnational corporations. These will, in turn, distribute it to Chagas sufferers. This is true of the two existing drugs developed by laboratories at Roche (Swiss) and Bayer (German).

The second aspect reveals the dynamics of the actors: Whereas scientists, doctors, and numerous state agencies publicly express their interests and capacities to mobilize social, discursive, material, and symbolic resources, affected people are in no position to flex a public voice. In the case of Chagas disease, this is, at least in part, because of its dual status as a "rural disease" which, therefore, affords little space for social interaction among sick people, being a "disease of the poor", whose symbolic—and, above all, discursive—capital is significantly lower when it comes to carrying an authoritative voice into the public sphere. So, where the affected people have no voice, other actors vie for the role of spokespersons, translating to the public arena the interests of patients, people at risk, and peasant and marginalized societies. By now, the paradox is obvious: It is precisely the lack of capacity of rural poor people

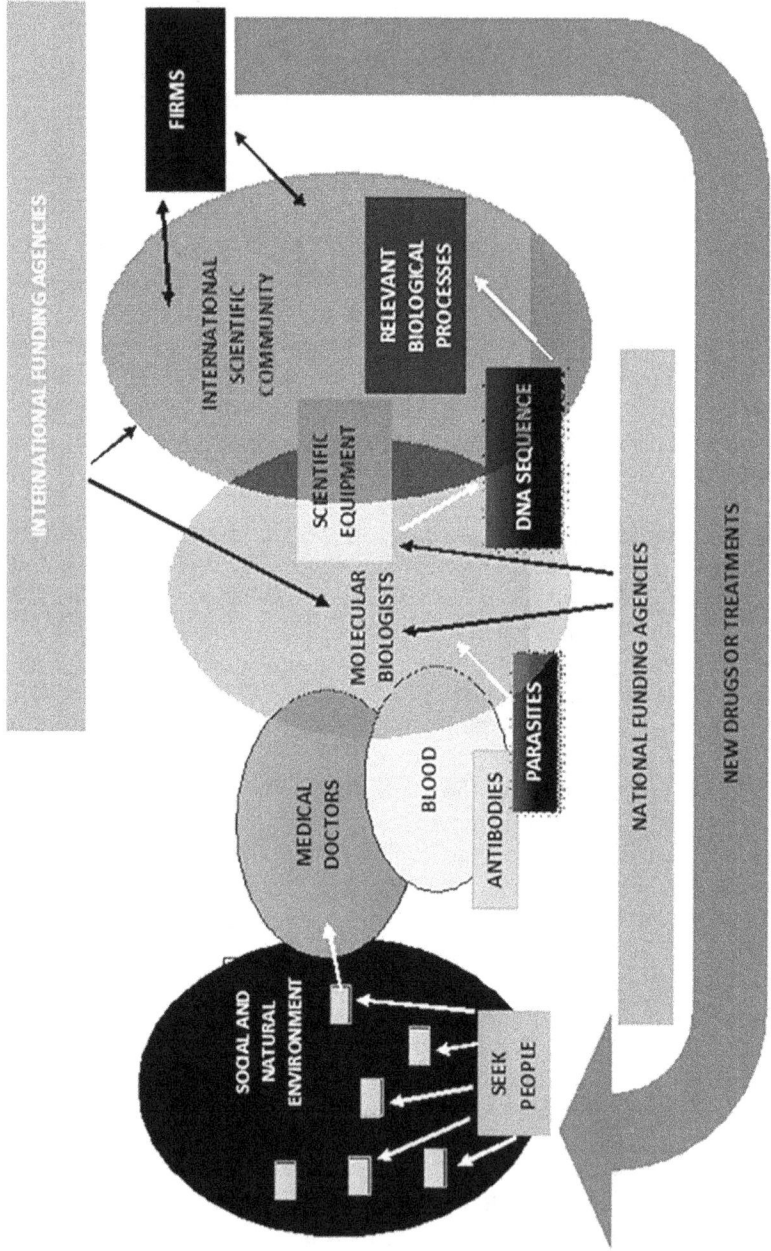

Figure 3.4 The Purification Process

that allows the development of multiple spokespersons, each bringing their own interests and their own construction to bear on the problem and trying to impose them as "objective things" in multiple fields of action. As is quite clear, the "purified" perspective we mentioned in the previous point has become a dominant perspective at least over the past fifteen years.

The third aspect refers to the dynamics of research and of knowledge production processes: the cognitive displacement from medicine (epidemiology, cardiology) through biochemistry and on to molecular biology also gradually reconstructed cognitive and social objects. This dual construction is important to take into account, because both dimensions are inseparable: if the parasite was first recognized as an "agent", this was because of its construction as an epistemic object, adopted and resignified by other social actors and thereby transformed into a social object with both symbolic and material existence. If, then, Chagas disease was associated with this parasite, this corresponded with a similar reconstruction process. The next shift from sick people to the disease and vice versa (enacted by various generations of medics, for example), which leads to Chagas cardiopathy, often independent of the rest of the body, is another of these processes. Later, the reconfiguration of the disease as *what happens to those living with parasite-transmitting vinchucas* and the set of actions that accompanied it (in particular, systematic fumigation) also involves a reconfiguration of the epistemic object. And, finally, the study of the parasite's DNA mentioned in the previous point again simultaneously reconfigures the cognitive and social objects through the purification process.

Thus, the fact that just 10 percent of research groups engage in clinical research—research that constructs knowledge by interacting with patients—and that more than 80 percent of research is located in Buenos Aires, where there is, of course, no vector transmission, are elements that can only be decoded in the framework of the dynamics I have discussed: namely, the mutual and simultaneous construction of the scientific and the social. I have started with a viewpoint from which the utility of scientific knowledge is seen in close connection with the processes involved in recognizing and shaping the social problem. In these processes, different conceptions of the central aspects of the disease and how to combat it are brought into play, as are the emergence of social actors with differing interests and the development of institutional frameworks in which these actors are embedded.

The articulations of various actors—researchers from different disciplines and fields, public authorities, physicians, patients, people at risk, and so on—influence the processes of production and use of scientific knowledge related to Chagas disease as a social problem and, conversely, the production and use of knowledge redefines the social problem, the perceptions of the actors, and ways of intervening in the disease.

On the basis of my empirical data, it is possible to organize the types of intervention in the disease. In effect, the production and use of knowledge have historically been taken as the *object* of research into the vector (the *vinchuca*), the causative agent (*T. cruzi*), or the host (the sufferer). The knowledge has also focused on environmental aspects, both housing and social, which intersect with the other approaches. A typology can thus be formulated, based on three "intervention modes", according to the *focus* with which the disease has been problematized. It will have been noted in the previous section that the same actors often appear in more than one configuration, while the mode they operate in differs substantially.

There is a significant shift in the center of interest, from *vinchucas* to parasites. The new momentum was grounded in the development of an important community of molecular biologists (in both Argentina and Brazil) "successfully" inserted in the international community through works on various aspects of the genetic devices and physiology of *T. cruzi*.

This phenomenon can be interpreted in terms of "subordinate integration" (a concept I proposed more than twenty years ago), which can be summarized as follows: Prestigious young researchers from peripheral countries often conduct studies in centers "of excellence" located in central contexts. There they are often given a line of research in response to an "agenda" arising from the needs of local labs and, naturally, of local societies. When these researchers return to their countries of origin, they often pursue the development of the lines of research they were working on during their stay abroad. As a result, researchers from peripheral countries continue to work on limited research topics, usually forming part of broader work programs thanks to their high technical skills (partly obtained in their training abroad). Their concentration on a narrow research line thus enables them to produce relevant, but "hyper-specialized", knowledge, which is why they often have no access to the general (theoretical) perspective of the problem.

Two general consequences in peripheral countries are particularly visible in Chagas research in Argentina: The type of subordinate integration strategy (predominant in the elite groups of scientific communities in peripheral countries) has a positive side, in terms of research groups, insofar as it makes it possible for peripheral teams to access international funding and other resources (equipment, and so on), which they would otherwise find it hard to gain access to.

The second consequence in this process is to observe that, insofar as local research groups design their agendas in close alignment with the dictates of the international scientific community, they generate portions (and products) of knowledge which, despite being "applied", do not have the capacity to generate effective applications or be appropriated by social actors external to the local scientific community. This phenomenon has been described as Applicable Knowledge Not Applied (AKNA), and its systematic character can be used as a true indicator of the structurally peripheral nature of local scientific culture, in particular, in the region's more scientifically developed countries, like Brazil, Mexico, or Argentina.[16]

We can see an international division of scientific work (a question I come back to in Chapter 7), where central laboratories usually impose research agendas both as a function of relations with the societies they are embedded in (with sturdy industrial fabrics) and of the most relevant topics of international scientific sub-communities. On the other hand, the scientists most integrated in peripheral contexts operate very often as true "subcontractors" of hegemonic groups (I will come back to this topic in Chapter 7).

In the case of Chagas research that I present, the most glaring evidence of this process is the vast number of articles published in prestigious international journals by local researchers. Not least, these researchers have published a fair number of these articles with "central" researchers or groups. Yet the development of drugs and other inputs to combat the disease has been practically nonexistent, despite the public rhetoric about the centrality of Chagas disease as social and scientific a problem.

Moreover, when an international organization, like the Drugs for Neglected Diseases initiative (DNDi), specifically designed to produce drugs "dropped" by international pharmaceutical laboratories, puts out calls to fund the development of drugs to treat Chagas (and two other diseases), only one of the more than fifty proposals received by the DNDi (twenty the first call, and thirty the second) came from a Latin American scientific research laboratory based in Venezuela.[17]

I opened the chapter by critiquing the widespread belief in the capacity of scientific knowledge to improve the living conditions of the Latin American population. We have seen, however, through a historical reconstruction of the emergence and development of a social problem (and its formulation in terms of a research object in a peripheral context) that a set of contingent and structural mechanisms limit these societies' capacity to take local advantage of the knowledge they themselves produce and finance. In general, the analysis shows us an interplay of actors and logics which, even though apparently convergent in their ways of defining and addressing the question, nevertheless conceal a dynamic that tends to perpetuate the social problem. This is because scientists seek their own legitimacy—and capacity for reproduction—in their specific fields, and therefore each actor turns any effective action on the disease into a rhetorical exercise. I take another (more enjoyable) look at this issue in the next chapter.

Notes

1. A structural weakness of most Latin American countries is the low quality of public decision-making, which is rarely supported by robust, locally produced knowledge. In contrast to most developed countries, Jasanoff (1998)

has developed the concept of "regulatory science" as a distinct domain of scientific production, accountable to epistemic as well as normative demands in ways that help explain why it is vulnerable to challenges from both science and politics. Indeed, regulatory science is extremely weak in Latin America, and very often knowledge produced in developed countries or international organizations (such as the emblematic case of WHO) tends to be taken as the basis for public decision-making. The subject does, of course, deserve a chapter of its own, but I want to at least mention it here.

2. I have addressed this topic in many texts, both conceptually in historical inquiry and also as a policy concern. See Kreimer and Thomas (2005) and Kreimer (2015). Here I prefer to tackle it using the concrete example of Chagas disease, with its potent heuristic capacity, for the reasons I explained in Chapter 1.

3. In the case of Chagas disease, one of the characteristics of the problem is precisely the lack of commercial interest in new medicines developed for pharmaceutical laboratories.

4. In other diseases, like AIDS, patients' associations have played an essential role in both allocating resources and defining desirable treatments. In the case of Chagas disease, patients' inability to organize can be seen as a consequence of the ignorance over their condition as Chagas sufferers, on the one hand, and of their determination to deny the condition in order to avoid the kind of discrimination it entails, on the other. For an analysis of the role of patients' associations as social actors, see Rabeharisoa and Callon (2008).

5. According to Bijker, "Relevant social groups do not simply see different aspects of one artifact. The meanings given by a relevant social group actually *constitute* the artifact. There are as many artifacts as there are relevant social groups; there is no artifact not constituted by a relevant social group". (Bijker, 1995, p. 77, author's emphasis). The concept of 'interpretative flexibility' was conceived to account for this multiplicity.

6. Among other honors, Chagas won the Schaudinn Prize in 1912, awarded every four years by the Bernhard Nocht Institute of Tropical Diseases in Hamburg for work in parasitology and tropical medicine. He was even nominated for the Nobel Prize in 1913 and 1920, though he never received the distinction. See Coutinho (1999).

7. At the Ninth Meeting of the Argentine Pathology Society in 1935, there were 35 cases; by 1939, the MEPRA had already reported 1,232 cases, while other foci were reported in Brazil.

8. For one thing, Carrillo was a native of Santiago del Estero, a province with high rates of endemicity. He had also practiced at the Surgical Clinic Institute run by Professor Arce (who pushed through the creation of the MEPRA), where he met Salvador Mazza.

9. The institute was so named to honor a young doctor who collaborated with Cerisola and died after accidentally becoming infected with *T. cruzi* in the laboratory.

10. Since this chapter was originally written ten years ago, the current structure of research focused on Chagas disease may differ, as it is certainly a very active field. For instance, several groups investigating vector populations emerged during the period. There were also deep changes in the distribution of *vinchuca* populations, due mostly to the extension of farming activity (closely associated with soy production) in wild, formerly uncultivated areas.

11. This topic is further developed in Chapter 5.

12. Parodi returned after a few years working at the Leloir Institute, in San Martín, where he has spent most of his scientific life.

13. In Argentina, the main researchers participating in the project belong to the groups of Carlos Frasch, a former researcher from the Campomar Foundation, and Mariano Levin from the INGEBI.

14. In 2008, the Institute was reorganized and changed its name: In 2008 it was transformed into Institute of Scientific and Technological Research for the Defense (CITEDEF) (Decree N° 1451/08)

15. The network created around this group includes teams from the University of Illinois, the Centers for Disease Prevention and Control (CDC) in the United States, the INGEBI, and the Fatala Chaben Institute in Argentina, the French Research Institute for Development (IRD), and the Argentine Ministry of Health's National Vector Control Program; and economic support from the University of Buenos Aires (UBA), the National Agency for Scientific and Technological Promotion (ANPCyT), the National Institutes of Health (NIH), and the National Science Foundation (NSF) in the United States.

16. I developed this concept many years ago (Kreimer & Thomas, 2005). Although the notion entails obvious epistemic problems (What is applicable? Who defines it and how? Is it something intrinsic to knowledge or does it depend on a network of social relationships for its meaning? and so on), it is nevertheless extremely useful when it comes to discussing science policy in developing countries.

17. Drugs for Neglected Diseases initiative (DNDi) is a collaborative, patients' needs–driven, nonprofit drug research and development (R&D) organization that is developing new treatments for neglected diseases. Originally focused on just four diseases, today it addresses the following pathologies: leishmaniasis, sleeping sickness (human African trypanosomiasis), Chagas disease, pediatric HIV, filarial diseases, mycetoma, hepatitis C, and malaria. www.dndi.org/about-dndi/ retrieved in September 2018.

4 Three Theoretical Divertimentos

Chagas Disease as a Model (Part 2)

In the previous chapter, I showed the process involved in the historical coproduction of Chagas disease as both a social and a scientific problem, and how the two areas have shaped each other.

In this chapter, I intend to look deeper into the same question, but from a different point of view: taking the form of an exercise or divertimento, I shall look at the "Chagas problem" from four well-known (and successful) sociological perspectives and make an effort of reflexivity and situate myself in different points of view. I also wish to persuade the (ever-skeptical, ever-leery) reader that there is no single way to formulate a scientific problem; rather, it depends on the disciplines or fields in question, levels of analysis selected, knowledge available, existing resources, or cultural, ideological, and other kinds of stimuli and barriers. Obviously, the same also applies to social issues.

In order to show how the mutual constructs between scientific knowledge and social actors operate in a complex and polymorphous way, I shall consider some episodes in the history of Chagas disease in Argentina with a view to questioning the process whereby it was constructed as a "public problem" in the course of the twentieth century, from the identification of the disease and its pathogen to its more recent reformulation by molecular biology.

Although the role of science as a "constructor" of social problems is not something completely new, the hypothesis can be formulated that there is today a new configuration of the role of scientific knowledge in the dynamics of formulation and strategies to solve social problems. From Joseph Gusfield's early 1980s book showing how alcoholism as an illness was a "pure construction" of scientists, to environmental issues and the controversy over genetically modified organisms (GMOs), science has consistently presented new problems in the public arena that are then *resignified*, *filtered*, and *processed* in different ways.

Starting Points

In a book more than two decades old, Isabelle Stengers (1997) proposed we consider three different social problems: the construction of a bridge, the construction of a dam, and the definition of "what is a drug". She analyzes the consequences of each in terms of knowledge, experts mobilized, sciences, and powers, emphasizing the plural that underpins these last two terms. Stengers dwells particularly on the consequences of the last two cases: the dam because, unlike the bridge, it interferes—or can interfere—in the course of the river; and drugs, because the saying "Tell me who defines it and I'll tell you what your problem is" seems to apply. Stengers's analysis is fairly close to other similar approaches proposed, for example, by Latour (1999) in his book on the politics of nature. The two books were published more or less simultaneously, and their subtitles are also very close: "La démocratie face à la technoscience" (Democracy faces technoscience) and "How to Bring the Sciences into Democracy". This is framed by a concern very common among our European colleagues toward the end of last century: the relationships between sciences (or technosciences), which have conquered enormous power and relative autonomy among the various public authorities.

As I noted in Chapter 3, the production of scientific knowledge has been actively formulated by public policies as a legitimate means of intervention in diverse social problems. This forms the basis on which science policies' legitimacy has largely been built.[1] In Latin America, since the 1960s, the state has developed instruments of intervention to promote and steer science as a "national" matter (Velho 2011; Feld 2015).

In line with certain fairly recent developments in the social studies of science, I wish here to break with the naturalistic narration of the social sciences. In fact, as John Law (2006) pointed out, even when we admit that social reality is complex, tortuous, and multidimensional, with numerous twists and turns and multiple simultaneous events, our explanations tend to be constructed on a linear narrative with a univocal temporal axis, relatively stable stages and environments, and so forth.

To counter this perspective, I imagine that, as in Akira Kurosawa's film *Rashomon* (1950), where different characters give their "real" versions of a case of rape and murder, we could counter a simple statement like "Chagas disease is a social problem" with a small set of interpretations so diverse that they can show us how far the interplay between the various ways of couching the issue are constructed and reconstructed, depending on the entrances and exits of the actors on stages which they, in turn, also change.

I will being this journey with a bald and—no doubt—provocative statement and then analyze it in its manifold aspects:

Chagas disease is not a social problem.

Shocked, any interlocutor sensitive to the rural poor's condition in Latin America would argue:

How can you say such a terrible thing! In Argentina alone, there are something like 3 million people infected with the *Trypanosoma cruzi* parasite (and around 20 million throughout the region). And they're all poor, with no access to early diagnosis, effective treatment, etc.!

In fact, our interlocutor is partly right: Chagas disease does affect several million people, not just in Argentina but in almost all Latin American countries. In fact, from Mexico to Patagonia, it is the only fully "Latin America disease". It is also a disease that cuts across social differences: the parasites are hosted by animals (and humans) and transported by certain insects (called vectors)—the *vinchuca* in Argentina, the *barbeiro* in Brazil, the *pito* in Colombia and Venezuela—which lodge in the nooks and crannies of *ranchos*, precarious rural dwellings of adobe and straw. There are no effective medications to treat it; for the acute phase, there is a single drug (benznidazole, produced by Roche), but it is more than forty years old, and production in Argentina is being phased out for market reasons (Kreimer & Corvalán 2010). There used to be another drug (nifurtimox, produced by Bayer), but production was stopped a few years ago. Additionally, the drug available has complicated effects, its application in chronic cases having been the subject of debate among the specialist community for quite some time, and no solution has been put forward until very recently. So, whether or not the chronic cases detected are treated with medication depends (or did so, at least, until a few years ago) only on chance: that is, on whichever line the treating physician prefers.[2] Let us just say, finally, that for many years there was no research undertaken by pharmaceutical laboratories— Latin American or transnational—to develop new drugs against Chagas, as evidenced by an international nongovernmental organization (NGO) dedicated to neglected diseases: the Drugs for Neglected Diseases initiative (DNDi).

Nevertheless, we tried to ask our indignant interlocutor to wait a little for us to show him different perspectives to help rethink the "Chagas issue", in the hope we can convince him that what looks like a social problem from a certain point of view (construct) may not look the same from another.

In the following sections, I take three very different theoretical conceptions focused on authors who have had an extraordinary influence on twentieth-century sociology: Bruno Latour and his theory of actants and allies; Theodor Adorno, who inquired into "legitimate" objects of sociology; and Pierre Bourdieu, who directs his attention at the fields of symbolic production and the problem of autonomy. From these three perspectives, I intend to show a mosaic—certainly incomplete and fragmentary—of multiple interpretations admitting the relationships between scientific knowledge and social problems. In the last section, I will reprise certain aspects and try to reorganize them, while introducing some new issues that—because of their scope—will have to be revisited in the future.

Act One: Networks, Allies, and Actants

The first exercise I propose is to consider some of Bruno Latour's ideas and see how we might apply them to the analysis of Chagas disease. Latour critically defined what he calls a "diffusion model" to explain relationships between science and society. He has stated that

> In the diffusion model society is made up of groups which have interests; these groups resist, accept or ignore both facts and machines, which have their own inertia. In consequence, we have science and technics on the one hand, and a society on the other.

This perspective is countered by a "translation model", which, according to him, has the advantage that there

> is no equivalent distinction, because there are only heterogeneous chains of associations which, from time to time, create obligatory passage points. Let us go further: the belief in the existence of a separate society of science and technology is the product of a diffusion model.
>
> (Latour 1989, p. 34)

So, we can assume there are three areas—science, technology, and society—that force us to study the impact of each on the other! Latour's solution to break with this perverse separation, which is a result of the incomplete development of the so-called modern constitution, is well known. On the one hand, he posits that there are no "pure" objects: everything we observe around us forms a proliferation of hybrid objects— hybrids by both nature and culture, that is (Latour 1991, p. 68). On the other hand, he proposes the category of *actants*, taken from linguistics. He claims that

> Since both humans endowed with words, and non-humans dumb, have spokespersons (who speak in their name), I propose to call "actants" all those, human or non-human, who are represented, in order to avoid the concept of "actor", overly anthropomorphic.
>
> (Latour 1989, p. 202)

Let us look at how we might analyze the emergence of Chagas by following these ideas, which can be précised as follows: Salvador Mazza, an Argentine doctor specializing in bacteriology, travels to Jujuy to study regional diseases and, after living there for a time, aims to identify the disease. His work sets out to identify Chagas disease as distinct from goiter, the two ailments being closely associated between the late 1920s and early 1930s, as Carlos Chagas asserts (a belief widespread at the time). A trip he made with Charles Nicolle to endemic areas in the northern provinces would later lead to the creation of the Mission for the Study of Argentine Regional Pathology (MEPRA), which, from 1933 on, focused on the study of Chagas. The history of the disease is interesting in one particular sense: Unlike most pathologies, where first the disease is established and then attempts are made to identify the causative agent, *Trypanosoma cruzi* was first discovered, and various diseases (such as malaria and goiter) were later attributed to it.[3]

We can then define our exercise as an attempt to draw a parallel between Latour's analysis (1983) of Pasteur and then to apply it to the development of Chagas as an object. Very briefly, Latour argues that to demonstrate the existence of microbes (in particular, the anthrax bacillus), Pasteur mobilized different actants, primarily farmers, moving his laboratory to the fields and selecting from them the elements of the natural world he will take back to his laboratory. To recruit their animals, he has to prove to the farmers that they will not die. The bacillus, as might have expected, is an actant represented by Pasteur, who also "translates" the farmers' interests. He thus persuades them to let him inject their pigs and goats with the bacilli in order to prevent them dying: in other words, he vaccinates them. Pasteur mobilizes the microbe as an ally to show military doctors that, if they boil water, these invisible beings he was telling them about but they cannot see will die and therefore, their troops will not become infected. In other words, he pasteurizes them. He does something similar with hygienists: He persuades them that communicable diseases are not "something that happens to bodies", but that "someone" does it to them. And he speaks on behalf of those invisible beings:

> How has Pasteur succeeded in capturing the interests of other indifferent groups? By the same method he has always used. He transfers himself and his laboratory into the midst of a world untouched by laboratory science. Beer, wine, vinegar, diseases of silk worms, antisepsy and later asepsy, had already been treated through these moves. Once more he does the same with a new problem: anthrax. The anthrax disease was said to be terrible for French cattle. This "terrible" character was "proven" by statistics to officials, veterinarians and farmers and their concerns were voiced by the many agricultural societies of the time.
>
> (Latour 1983, p. 145)

The notion of interest is key in Latour's explanation. For him, "interests, like everything else, can be constructed". The processes involved in constructing interests pass through the mechanism of translation: It is through the operations of translation that an actor brings the interests of others into play in the hope of imposing their—the actor's—own meaning on what is "at stake". However,

> The translation that allows Pasteur to transfer the anthrax disease to his laboratory in Paris is not a literal, word-for-word translation. He takes only one element with him, the micro-organism, and not the whole farm, the smell, the cows, the willows along the pond or the farmer's pretty daughter. With the microbe, however, he also draws along with him the now interested agricultural societies. Why? Because having designated the micro-organism as the living and pertinent cause, he can now reformulate farmers' interests in a new way: if you wish to solve *your* anthrax problem you have to pass through my laboratory first.
>
> (Latour 1983, p. 151)

The analytical categories of hybrids and actants are the result of the concept of extended symmetry, where no analytical distinction must be made, *a priori*, between the natural and social worlds, both being, in practice, inseparable. These categories were—and still are—annoying to sociologists, who need to observe social relations, social facts, and discourses produced by subjects, among other configurations and devices. But behind each, there are social subjects whose practices cannot be reduced to the one-dimensional aspect implicit in Latour's theoretical proposal. In fact, the development of knowledge is explained there by the interests of specific subjects and of the processes of enrollment/mobilization/translation that operate with other actors/actants. There is no place in this perspective for the cultures of the subjects and groups in question, nor for the level of institutions, to mention the two most obvious limitations.

Following Latour's methods, we can say that Mazza himself tried to speak on behalf of *T. cruzi* when he decided to move to Jujuy to study Chagas patients and develop methods to identify the disease. Like Pasteur, he needed to persuade some very powerful actors, like the medical elites in Buenos Aires (the Faculty of Medicine of the University of Buenos Aires [UBA]) which, while it encouraged him at first, later raised objections and eventually withdrew its support. He also had to convince his Brazilian counterparts, who, under the influence of Carlos Chagas Filho, had yet not accepted the connection between the parasite and disease. To achieve this, as well as establishing himself as the parasite's spokesperson, he needed to persuade the population—the patients—that they were infected with an invisible being and that he was able to detect it: He had to get them on his side, in other words. So when Cecilio Romaña claimed to have developed an indicator/symptom to diagnose the disease (an ocular edema that, given the then fashion for eponyms, came to be christened "Romaña syndrome"), Mazza fought him tooth and claw, needing a monopoly on the social representation of the parasite in order to legitimize his own position. The ending is well known: Mazza was betrayed, so to speak, by his own support for the medical community and by the very parasite he studied. He contracted the disease himself, and this, paradoxically, was perhaps his greatest triumph.

At any rate, in the origins of Chagas disease, the "problem" is defined in terms of a "new disease", for which Mazza becomes the spokesperson, publicly suggesting the idea that the issue can be addressed by taking the *ranchos*, where the insects that transmit the parasite are found, as the object. His operation has two phases, the first successful, when he manages, by setting up the MEPRA in Jujuy Province (in the northwest of Argentina, the heart of one of the disease's endemic areas), to coordinate a network that has him as both axis and intermediary. He is able to determine the presence of parasites in the organism and diagnose people who are sick and people who are not, and what proportion of the rural population is infected, through the analysis of droplets of blood. In a second phase, Mazza gradually loses his allies at the UBA, makes enemies with some of his students, and decides to move the MEPRA to Buenos Aires, where his role as exclusive mediator with *vinchucas* and parasites weakens to a fatal extent. The "Chagas issue" comes to have less public visibility, until it resurfaces some time later, reformulated as a "national problem" of health, by Mazza's student, Cecilio Romaña, who distances himself from his master and plays a crucial role during the Peronist period (1945–1955), setting up a totally new network of allies when compared to the strategy and interests mobilized by Mazza.

Act Two: "Essential Problems"

Now let us look at a contrasting perspective and Latour's response:

> An objection frequently made to the concept of society today is that it is a metaphysical concept. It is very interesting—and it is a piece of modern ideological doctrine of which I should like to make you aware—that critical ideas are no longer attacked, as used to be the case, as corrosive or aggressive or in suchlike terms. Instead, the attempt is made to dispose of them by saying that they have fallen behind current developments, and that any view which does not accept the existing order is a kind of residue of ancient metaphysics, ontology, or the disguised theology for which I am criticized by Scheuch, or whatever else. Ladies and Gentlemen, the fact that this kind of apologetics is predominant today throws light on the general state of society.
>
> (Adorno 1968, p. 28)

The quotation is by Theodor Adorno and is from 1968! Next, Adorno mentions what, for him, is sociology's field of study. On the one hand, he points out that

> That is to say, there is nothing under the sun, and I mean absolutely nothing, which, in being mediated through human intelligence and human thought, is not also socially mediated. For human intelligence is not something given to the single human being once and for all. Intelligence and thought are imbued with the history of the whole species and, one may also say, with the whole of society.
>
> (Adorno 1968, p. 30)

He also points out that "sociology's interest should be directed at the essential, that it should deal with socially relevant matters and not with subjects which are of no interest". Adorno immediately warns that the essential cannot be identical to the "big issues", especially because "one cannot see straight away, just by looking at an object, whether it is essential or not. As a rule this is decided only in the execution, by what is revealed to us through the object" (1968, p. 33). In other words, it is the very approach to objects that constructs their meanings, and those that appear to be divergent and opaque phenomena can, many times, lead to extremely relevant social perspectives.

Thus far Adorno is clear. We suppose that, if we were dealing with Chagas disease as a "marginal" problem, Adorno might have laid bare the frameworks of the devices that produce social exclusion. The supposed naturalization of a "mere health problem"—note the derogatory tone—would thus hide more relevant issues: namely, in this case, the consequences of deeply dual rural production modes like the high productivity of the pampas (a hegemonic model of production), on the one hand, and the precapitalist, self-consumption, or simply marginal modes of production specific to areas where Chagas disease is endemic, on the other.

But in terms of the construction of sociology's objects, this epistemological perspective has its limits: To what extent do *real problems* seem to be beyond reasonable doubt? In Adorno's texts we find an approximation to this concern:

> Please excuse me if I give a crude example, which I choose only to make clear to you something which easily escapes our awareness. It is that decisive discoveries in medicine, such as that of the cause of cancer and therefore a possible cure for cancer, would probably have been made long ago had not a wholly excessive amount of the social product been spent, for social reasons, on armaments or the exploration of empty stars for advertising purposes [. . .] But it seems absurd to me that such elementary needs and problems, which affect human

life as directly as the possible cure of allegedly incurable illnesses—which, I have been told by various doctors, could in principle be cured—remain unsolved for social reasons.

(Adorno 1968, p. 30)

The example is, of course, hard to reject and refers to the exceptional nature of certain issues, as the practical always creeps into this question about "the essential". Therefore, in any theory of society, certain issues directed at the subjective, which have no great importance in and of themselves compared to the structural problems of society, do, nevertheless, possess dignity. For Adorno,

> They are important because—and I cannot help saying this—after Auschwitz (and in this respect Auschwitz is a prototype of something which has been repeated incessantly in the world since then) our interest in ensuring that this should never occur again—or, where and when it occurs, that it should be stopped—this interest ought to determine our choice of epistemological methods and our choice of subjects to be studied [. . .] It may be that the murder of six million innocent people for a delusory reason is an epiphenomenon when measured by the standard of a theory of society, something secondary which is not the key to understanding. However, I would think that merely the dimension of horror attached to such an event gives it an importance which justifies the pragmatic demand that in this case knowledge should be prioritized—if I may use that dreadful word—with the aim of preventing such events.

(1968, p. 33)

Faced with this crucial boundary, our new obstacle arises again, the "absolute evil" of Auschwitz aside, the horror of which seems beyond all analysis: Where to draw limits between the essential problems and others, which Adorno terms "practical"? In fact, rather than being practical, our problem directly involves the role of the sociologist and the type of knowledge that sociological practices generate, or should generate (for the horror not to be repeated, for example). One might argue that, under the Nazi regime, it is not about the role of the sociologist but about the citizen, the social subject pure and simple; but to this Adorno adds "to promote knowledge so that this is not repeated", and that is already a task of sociology.

Let us come back to our example of Chagas disease. How far is the extent of the disease a "limit situation" therefore requiring a "pragmatic" solution? Or is it a "resulting epiphenomenon" and, therefore, a matter that should be set aside in order to investigate the true social roots of the problem? In one alternative, if we take into account the scale of the disease, the (sometimes extreme) state of poverty, and the other conditions of destitution, the answer is immediate: We cannot tolerate such living conditions *without producing the knowledge to prevent this from happening*. On the other hand, if we are interested in the "*structural problems of society*", we must ignore it out of hand, or, in other words, not be led by apparent situations of social gravity and deal instead with the *real problems*.

In Adorno's sense, the very existence of Chagas as a social problem is at issue. Let us look at what happens if we return to our initial statement, "Chagas disease is not a social problem":

1. While it is true that existing records speak of between 2.5 and 3 million people infected in Argentina, these records are very unreliable: The latest aggregated data are from the last lists of military service conscripts, in 1995.[4] Two remarks in this regard: The first that, after several years, and, in particular, after the deep economic and social crisis of 2001–2002, it is not possible to assume that the trends then witnessed have anything more than an approximate value today. Chagas being a disease almost exclusively affecting the rural poor, it is plausible to assume that it is highly dependent on the overall economic conditions affecting those sectors. The second remark alludes to a gender issue: Military service was compulsory only for males, so there has never been a record of the disease's incidence in women. Here, then, we see a fiction: The parasite is "supposed" to affect both sexes—and therefore steps are taken—given that there is no study to make us think this is so.

2. As we have seen, Chagas almost exclusively affects poor rural sectors. But these sectors live in extremely precarious conditions in terms of facilities of drinking water, sewers, and availability of drugs, in addition to serious nutritional deficiencies and so forth. Whereas women's life expectancy at birth in Buenos Aires in 2001 was 79.39 years, men's in Chaco (a northeastern province close to Paraguayan border) was 66.95 years, almost thirteen years less.[5] We must add that, whereas there is no rural population in Buenos Aires, the Chaco figures do include its urban population, raising the indicator by several years. In other words, a significant portion of Chagas patients will not die of the disease but will be affected by other ailments and, in particular, by their wretched living conditions. If to this we add the fact that the development of the indeterminate stage of the disease may take more than twenty or twenty-five years, we can conclude that many of those infected will never even find out they are suffering the ailment because they will never develop the chronic phase.

3. The very people affected tend to *naturalize the existence of disease* when officials from the Ministry of Health—national or provincial—visit populations at risk and ask them what diseases are prevalent there. The most frequent responses are often a short list of ills like

measles, chickenpox, flu, and so on. When the official asks them "No Chagas, right?" the answer many times is "Ahh . . . yes . . . Chagas, everyone has Chagas here" (Sanmartino & Crocco 2000).

4. Except in its acute phase (for which the drug treatment is quite effective) and in the later development of heart disease, the disease is not in the least disabling. On the other hand, only about 20 percent of infected people who later go on to the indeterminate period will develop the chronic stage, more than twenty years later, which is accompanied by more severe lesions. And, as we have seen, many of them will suffer from other conditions before that. In other words, for 80 percent of the population infected with the parasite, Chagas is not a subject they need to worry about. This last argument could be refuted by replying that, nonetheless, it cannot be known in advance who will belong to the 20 percent that will develop the disease in the long term. This is true, but that does not invalidate the fact that the chances of doing so remain statistically low.

Having reached this point, we now have arguments to rethink the existence or nonexistence of a social problem in terms of other knowledge quite different from that used by the state to propose the mobilization of scientific knowledge as a legitimate mechanism to address the problem.

Act Three: Autonomy and False Social Demand

The well-known works of Pierre Bourdieu are another vantage point from which to tackle our issue. Indeed, Bourdieu has dealt with the issue of scientific knowledge, its social organizations, and its relationship with society, where fields are framed in various moments of his intellectual development.[6]

In fact, right from his first text on the scientific field in the 1970s, Bourdieu drew attention to the problem of autonomy as one of its constituent aspects:

> Strategies of false separation express the objective truth of fields which have only a false autonomy: whereas the dominant class grants the natural sciences an autonomy corresponding to the interest it finds in the economic applications of scientific techniques, so that they are now (even for the religious consciousness) fully autotomized in relation to the laws of the social world, the dominant class has no reason to expect anything from the social sciences—beyond, at best, a particularly valuable contribution to the legitimation of the established order and a strengthening of the arsenal of symbolic instruments of domination.
>
> (Bourdieu 1975, pp. 35–36)

From then on, Bourdieu would become not just an analyst of fields of symbolic production but a fighter for more autonomy. For him, "the more autonomous one is, the more chances of wielding specific authority (scientific or literary) that authorizes one to speak outside of the field with a degree of symbolic efficacy" (Bourdieu 1997, p. 65).

In his last writings, the problem of the autonomy of the scientific field—or rather, the threat to it—emerges in what we can only call dramatic fashion. In his last text on the subject, Bourdieu was horrified:

> The autonomy that science had gradually won against the religious, political or even powers economic, and, partially at least, against the state bureaucracies which ensured the minimum conditions for its independence, has been greatly weakened. [. . .] In short, science is in danger, and for that reason it is becoming dangerous.
>
> (Bourdieu 2001, p. 7)

It is, then, from this strong belief in the notion of autonomy as the precondition for the field's effective operation (in the sense that its internal logics of legitimation of knowledge prevail over external logics, which are perceived as "imposed") that Bourdieu introduces the problem of the "social demand" for knowledge. Faced with this stance, he pulls no punches: Such a pretended demand is nothing but a euphemism to conceal specific interests which, almost by definition, fall far short of meeting the real needs of the social agents who actually suffer them. For Bourdieu,

> There is every reason to think that the pressures of the economy are growing more intense with each day that passes, especially in areas where the results of research are highly profitable, such as medicine, biotechnology (in agriculture in particular) and, more generally, genetics—not to mention military research.
>
> (Bourdieu 2001, p. 8)

Against this background, according to Bourdieu, selfless scientists tend to disappear, as they know no other program than the one to be gleaned from the logic of their research and know how to give "commercial" demands the strict minimum of concessions essential to ensure the credits necessary for their work.[7]

In one of his true diatribes at researchers who asked him what to do then against social demands, Bourdieu replies that it is necessary, above all, to modify those scientists' "mental habits" who only feel universal when defending interests not their own and set themselves up as carriers of a "social demand". Instead, he says,

> I believe (you) should begin by asserting your autonomy, by defending your specific interests, that is to say, in the case of scientists, the

conditions of scientificity, and so on, and from there to intervene *in the name of the universal principles of your existence* and the conquests of your work.

(Bourdieu 1997, p. 67. My emphasis.)

Here, something already hinted at in a previous quotation appears for the second time: The defense of autonomy is a prerequisite for the scientist, the intellectual, the *savant* themselves to intervene outside their own field. He has in mind—and says so—Emile Zola and the Dreyfuss Affair: He intervenes in a public issue as a prestigious intellectual, but, having done so, returns to the literary or intellectual field. So, in Bourdieu's view, autonomy runs in a single direction: the protection of the mechanisms that guarantee scientificity—assuming they are not constructed—from all external (political, social, economic) interference, but, at the same time, he allows—encourages!—scientists' intervention in other fields. This proposal is strikingly asymmetric: While other actors external to the scientific field have to be kept out of it in order to preserve autonomy, scientists, once their symbolic capital has been secured, are invited to intervene in other fields *in the name of that accumulated capital*.

Bourdieu's perspective has a further drawback: By equating "social demand" with the "businesses' use for financial gain", he dramatically loses sight of scientific knowledge's role in society. Scientific knowledge is always the bearer of a dual legitimacy: the ability to explain the workings of the physical, natural, and social world, on the one hand, and, on the other, the ability to transform those worlds to satisfy the needs, issues, or demands of different actors in society. Therefore, the first, cognitive dimension is inseparable from the social dimension, the circuits of knowledge as a product, "what do we do with it", and so forth. But knowledge can never be used by another actor "as is" (no one is cured or fed or produces more because of a paper); this can only be done through a complex process of transformation and resignification of knowledge. To simplify, we can call this process "knowledge industrialization", involving both end and intermediate users, who are precisely those capable of industrializing knowledge.

At this point, we are now in a position to return to our exercise and imagine how the construction of Chagas disease as a social problem might be analyzed from Bourdieu's perspective.

First, we have Dr. Salvador Mazza, a student at the National School of Buenos Aires and trained in the UBA's Faculty of Medicine at the start of the twentieth century who was then working in microbiology, an emerging discipline that quickly spread far and wide after Pasteur's revolution. Mazza accumulated significant symbolic capital: In 1916, in the middle of World War I, the Argentine army commissioned him to carry out a study of infectious diseases in Germany and Austria-Hungary; at that time, he met his colleague Carlos Chagas, who had recently identified *T.*

cruzi. On his return to the country, in 1920, he was appointed director of the central laboratory at the National Clinics Hospital and director of the Bacteriology Department in the Faculty of Medicine. In 1923, he headed for France to further refine his studies and arrived at the Pasteur Institute in Tunisia (still a French colony), directed by the bacteriologist and entomologist Charles Nicolle (awarded the Nobel Prize for Medicine in 1928). Mazza returned to Argentina in 1925 and was appointed head of the laboratory and museum of the Surgical Clinic Institute in the Faculty of Medicine. At the end of 1925, he invited Nicolle to Argentina and gave him accommodations, as the Frenchman was interested in the endemic diseases in the north of the country.

The support of Dr. José Arce, a medical doctor, former deputy, and later rector of the UBA, was crucial in the creation of the MEPRA, and Mazza, using the symbolic capital obtained in the scientific field, decided to intervene in defining what was *scientifically relevant*. He would, over the years, define as a central cognitive problem how to determine the connection between *T. cruzi* and Chagas disease independently of endemic goiter, and would concern himself with developing and proposing different techniques for diagnosis and identification of infected people. On the basis of this capacity, Mazza managed to deepen his strategy to establish a subfield that would autonomize itself from the political power that gave rise to it and the increasingly systematic detection of cases would allow him to show the relationship between the living conditions of infected people and the spread of the disease in endemic areas. This is also seen in his capacity to establish the three phases that characterize the disease: acute, chronic, and indeterminate. During that period, which lasted up to the 1940s, Mazza clearly seems to obey Bourdieu's precept:

> Why should writers and experts not participate in the definition of the social demand? Armed with the achievement of work and expertise possessed by scientists, they could intervene effectively in problems of general interest, and not only intermittently, when politicians cross the line, but customarily.
>
> (Bourdieu 1997, p. 60)

During those years, in addition to "publicly imposing the meaning" of Chagas disease, Salvador Mazza did not forget to develop the legitimation mechanisms of the scientific field themselves: Through the MEPRA, more than 300 publications were produced, and growing numbers of papers were presented at disciplinary conferences.

However, the growing involvement of political agents and scientist-competitors from the field itself was affecting/weakening the capital accumulated by Mazza—and the autonomy he had managed to establish—and he had great difficulty keeping the MEPRA up and running in his last years, although the institution survived until 1958, a full fourteen years.

In this analysis, Mazza threw out any problem related to the false existence of a "social demand" in order to concentrate on his own dynamics within the field, which was what allowed him to intervene successfully outside of it. In other words, Chagas disease, its relationship with the parasite that causes it and the *vinchuca* that transmits it are, in fact, real issues formulated by Mazza thanks to the symbolic capital and autonomy he enjoyed and, therefore, to the absence of interested (political or economic) interventions attempting to steer his research.

Act Four: Knowledge Constructs Public Problems

As is often the case, after critically presenting—in the form of an imaginary exercise—the three highly prestigious sociological perspectives whereby the emergence of Chagas might be analyzed as a social problem, I shall set out a different way to analyze this topic, demonstrating, at the same time, the social dimensions that construct the problem and the role of knowledge in these processes.

I shall take as a starting point a work by Joseph Gusfield, who, in a book published in the early 1980s, studied the emergence of the dangerous relationship of drinking and driving in terms of the construction of a "public problem". The first thing Gusfield did was to denaturalize—and so deobjectivize—any interpretation presented as "given":

> Alcohol has already been perceived as important in the genesis of such fatalities [car accidents] and accorded an importance as target in the resolution of the problem. That target character is no a given, is not the nature of reality as a *Ding as sich* (a thing in itself), but represents a selective process from among a multiplicity of possible and potential realities which can be seen as affecting auto fatalities and injuries.
>
> (Gusfield 1981, p. 3. Author's emphasis)

To analyze the reasons why "something" attains the status of a public issue, the author first distinguishes private and public problems, taking into account that not every "social problem" will become a public problem, taking as such those that become a matter of conflict or controversy in the arena of public action.

Who or what institution has or has been given the responsibility to "do something" about the issue? Insofar as the phenomena are open to various ways of being conceptualized as problems, their public nature is also open to different ways of conceiving their solution.[8] For Gusfield:

> Science, scientific pronouncements, technical programs, and technologies appear as supports to authority, and counterauthority, by giving to a program or policy the cast of being validated in nature,

grounded in a neutral process by a method that assures both certainty and accuracy.

(Gusfield 1981, p. 28)

According to Gusfield, this is not about the "natural" resource that the actors make of the relevant knowledge produced by scientific and technical devices, but rather about certain actors making a deliberate and specific use of scientific knowledge as a way of mediating in public controversies over part of a problem that, precisely through these means, becomes public. In other words, it is no longer a matter of "science", but the role scientific rhetoric plays in the construction of public problems.

When he analyzes the way car accidents were constructed in a "fact" (in the Durkheimian sense), Gusfield draws attention to the fact that data are not simple facts collected by individual agents, but that these data are gathered and presented; the "discovery" of public facts is therefore a process of social organization: "Someone must engage in monitoring, recording, aggregating, analyzing, and transmitting the separate and individual events into the public reality of 'auto accidents and deaths'" (1981, p. 37). But he asks himself: What facts are aggregated? By whom? How are they processed? How are they transmitted?

Monitoring these questions allows the author to identify the corpus of research that forms the basis of drunk-driving policies. He observes here that there are two types of fictions in scientific analysis: The first is conceptual and refers to the treatment of theoretical entities as if they had real existence—and operate on them; this, for example, is what physicists and chemists do with concepts such as attraction, repulsion, and pressure, or sociologists with categories such as community, society, or capitalism. The second type of fiction is the "as if": This model assumes that a methodology might be applied that is aimed at dealing with confusing data and events "as if" they were similar to the phenomena of a known, familiar reality, a process that produces an illusion of certainty, clarity, feasibility, and authority.

If we apply an analysis analogous to that proposed by Gusfield to the process of constructing Chagas as a public problem, we should begin by asking ourselves how knowledge about the disease generated in the different actors for it to leave the private space of a group of infected people living in poverty in rural areas and take the public stage. In this case, it is possible to identify two very real entities—the *vinchuca* and the *T. cruzi* parasite—that are purified in successive laboratories: first by the medical doctors who, like Mazza himself, studied it in the field; then by epidemiologists, who "mapped" (what is more a part of the world of fiction than a map, clearly shown in Jorge Luis Borges's ironic text on the rigor of science?!) *vinchuca* populations. They were followed by sanitation agents and, associated with them, the chemists who produced the insecticides for two transnational laboratories (first the German Bayer, then the Swiss Roche) and conducted the first clinical trials to measure the effects of new drugs to kill the parasite. Last came the molecular biologists, who turned around all the preceding schemas, promising, toward the end of the 1970s and early 1980s, the development of a vaccine.

The political authorities struck a balance between the different areas of the state. On the one hand, they consider systematic fumigation a priority option to eliminate *vinchucas* and, therefore, "vector" transmission (there were numerous ventures, from the most vertical to the most horizontal, decentralized, and participatory policies). On the other hand, there was stimulated basic knowledge production, which one day promised to achieve the long-awaited vaccine.

This process can be illustrated in an example: Whereas the *golden age* of the promised vaccine performed like clockwork well into the 1990s, various groups benefited from the public funding of a problem defined in such terms that its solution appeared to be only "a matter of time". Other groups, for their part, worked on the construction—again rhetorical—of "targets" to attack the parasite and, therefore, that were the source of new drugs, as the two that were available had—and still do have, as we stated earlier—complicated side effects, and there was no full agreement on their effectiveness for the phase of chronic infection.

For a few years now, the once many research groups on the development of a vaccine have dwindled to just a couple. Slowly, the political authorities have gradually abandoned this strategy of dealing with the problem in order to prioritize two of the available solutions: systematicity in the control of the vector, as in the case of various Brazilian states (some of which have already declared themselves "Chagas-free") and basic research on targets to attack the parasite.

However, one group still researching the hypothesis of a vaccine claims to have one that is fairly well developed, based on attenuated, less virulent strains of the parasite. On top of the position over how to tackle the problem being marginal nowadays, other changes also affect this type of strategy: Clinical trials of new treatments, drugs, or vaccines are bound by many requirements, meaning that this vaccine, however effective it may seem, continues to operate as a true "as if" fiction (according to Gusfield). According to the head of the research group, "with today's clinical and ethical criteria, it would be impossible to test the vaccine in humans". So his only hope is "a violent outbreak of infected people in Bolivia, for example, and that, in the resulting health emergency", he and his team are called on "to go there to perform mass vaccinations and face even greater dangers". What is explicit in these plans exempts me from making any further comment on the subject.

Reassembling and Concluding

If I have mobilized various approaches, taking as a pretext what, *a priori*, would seem to be a "social problem beyond any discussion" (Chagas disease), it is to pursue my exercise of changing the "viewpoints" of the supposed problem.

So, if today we talk of Chagas disease—as we have been doing for decades—under the widely used label of "principal national endemic", this is due to the several constructions developed by a handful of actors.[9] We can assume that the use of scientific rhetoric has, among other consequences, the result that the actors who actually participated in defining "what is at stake" have spoken—and are still speaking—on behalf of the social sectors affected. But their representations are far from being neutral operations; on the contrary, they involve the deployment of symbolic devices in which, as in the theater of politics, the actors play a role and shape their discourses in the terms of that participation. All the mechanisms of rhetoric are brought into play; the only forbidden thing is to expose the connections between each actor's interests and stances in relation to the discourse that molds them. For example, a molecular biologist may refer to the possibility of using their knowledge for the production of a vaccine that will prevent x cases of the disease, drawing attention to the scarcity of such products, the severity and extent of disease, and so on. But what they cannot say in their public intervention is that it will allow them to successfully take part in an international network that studies regulatory mechanisms controlling gene expression. Thanks to this participation, they will obtain funding that will make it possible to expand their laboratory's equipment and technical devices, attend international conferences, build up their symbolic capital in academic institutions, and so forth. In the same way, a politician looking for votes cannot, for example, say that he or she needs to make "strong" statements in order to get the media's attention, because his or her advisors have pointed out it is the best strategy with the polls showing them to be an unknown candidate who has to "establish" their image.[10]

Other important aspects in the processes constructing social and knowledge problems emerge from the analysis I propose. If we say that scientific knowledge builds a social problem, we can take it as "objective" and the problem as "constructed", which conceals the constructed nature of the knowledge. The sociology of knowledge can make certain significant contributions in this direction, in particular, highlighting what we might call the "purification illusion": when scientists are studying the parasite—our issue at hand—they take that entity to laboratories and resignify it or construct it as a cognitive entity. In fact, seen in this light, the parasite is no longer *T. cruzi*, but a purification that has been decontextualized in test tubes and other devices but that acts as a "representative" of the natural world. As Karin Knorr Cetina pointed out, when one enters laboratories, nature is never to be found there, but a set of selections operated by the researchers. Thus, scientific objects are not only "technically" manufactured in laboratories, but inextricably constructed, symbolically and politically (Knorr Cetina 1995, p. 143). The issue leads us back to the fictions constructed in the public sphere: *Vinchucas* and parasites are presented "outside" of their significant habitat as a function of the disease in question, creating the illusion of purification and, therefore, material and symbolic command of the problem.

Another key aspect to emerge from the analysis I present here is that it ignores knowledge industrialization processes. It operates "as if" by asserting that work is being done on the production of a drug, but without the other elements that would allow its development to be actually present throughout the manufacturing process of the objects in question. I like to name this type of statement *magic realism*. For example, a target is defined, a way of attacking the parasite in order to kill it or prevent it reproducing in humans; then there is a search for the molecule that can do the job. Very often the research stops there. But, by that stage, there is a set of questions that cannot be answered, without which the knowledge produced is meaningless: Will the molecule only have an effect on the parasite or will it also affect other tissues and vital functions? If the parasite can be affected, is it possible to administer it to humans in a reasonable form (capsule, injection, syrup, and so on)? If several million doses per year were produced, are the technical means and resources available? And the two most crucial: at what cost, and who can manufacture and distribute it?

As we can see, without this set of questions (and answers), the issue of "targets" makes no sense. This is because, as I affirm here, these aspects belong to the content of the knowledge and are not ex-post additions to the "purified" process of research.[11]

Let us come back for a moment to the problem of autonomy: In more recent times—that is, with the emergence of molecular biology as a disciplinary field that reorganized the whole of the "life sciences"—the study of the parasite assembled some of the most prestigious groups in Argentina from the "hegemonic" tradition founded roughly mid-century by Leloir (Kreimer 2010d). Indeed, molecular biologists began to use *T. cruzi* as an analytical model with a manageable scale where they could identify a set of biological phenomena of international interest for the field. They had a "bargaining chip" with their peers in the central countries: While they were able to provide an exotic "bug" and analyze problems like the regulation of gene expression, they could successfully take part in partnerships and then integrate into "networks of excellence". They can provide specific knowledge to these networks and receive, in exchange, visibility, recognition, and technical, social, and cognitive resources. Leaning extensively on Bourdieu's recommendation, these molecular biologists powerfully asserted their symbolic capital and successfully positioned themselves as leaders in the field at the local level. Their output was significant, as measured—in classical terms—in articles in leading international journals (Kreimer & Zabala 2006, 2007).

At the same time, the problems arising from social demand were ignored insofar as, over the last quarter of a century, they have produced no therapeutic innovations, drugs, or vaccines to improve the condition or diagnosis of Chagas patients. Is it possible to say here, with Bourdieu, that scientists have "dissolved a false dilemma" insofar as they have concentrated on making extensive use of the autonomy they have enjoyed? If so, at what cost?

The answer is simple: They built *other problems*, because their activity is never "neutral" and not just seen within the field's boundaries, but, above all, through their role in the society that funds them. In terms of the construction of a public problem, their intervention has been decisive, having positioned knowledge production about the parasite center stage, at least partially, displacing other solutions or "selections" available to the actors, such as the systematic spraying of *ranchos*, among others whose "effectiveness" in fighting the social problem, as formulated, was apparent wherever applied. We therefore reach a paradox: The mobilization of scientific knowledge in tackling (or solving) a social problem depends on how the problem has already been formulated . . . by knowledge itself!

I would like to close this chapter with a personal anecdote. Some years ago, I attended a workshop organized by the TDR Program (belonging to the WHO), which discussed the current advances and perspectives in Chagas research and treatment. Several groups were represented: molecular biologists, parasitologists, medical doctors (especially cardiologists), epidemiologists, entomologists, blood bank managers, and so forth. The mutual unintelligibility of the languages used by each group quickly became evident: When one molecular biologist presented a model of apoptosis (the self-programmed death of a cell) in *T. cruzi*, I asked the cardiologist and the entomologist sitting next to me for an explanation (the subject sounded too complex for a sociologist), and they both confessed that they found it incomprehensible too. In contrast, when the cardiologists began talking about their experiences in the treatment of sick people and the changes observed in the development of the disease (the "urbanization" of Chagas due to rural–urban migration, for example, or increased intrauterine infection), the molecular biologists and biochemists made their boredom abundantly apparent. Each group had its own strategy for tackling the disease: vector control, epidemic surveillance, molecular research, clinical trials, and so on. On the basis of the apparent maturity of the different Chagas research fields, the WHO authorities proposed a change of category, from Type II (a disease with some breakthroughs but still needing further R&D) to Type I (a disease whose scientific basis is well known and efforts therefore need to be directed at control and treatment) (WHO/TDR 2005).

General pandemonium broke out. Despite disagreeing on almost every topic, all the participants at least agreed on one paradoxical point: "they have made crucial advances in the fight against this endemic disease", but "research into Chagas disease was not 'mature' enough, and more work was needed". Or to put it another way, "We need more funding to continue our important work".

Notes

1. Stengers statement of this problem is provocative: " 'It is proved that . . .', 'from a scientific point of view . . .', 'objectively, the facts show that . . .' How many times do these expressions chime in the discourses of those who govern us? Because, since our societies have striven to be democratic, the sole authoritative argument about what is possible and what is not has come from science". Stengers (1997, p. 2)
2. I analyze this controversy in Chapter 6.
3. For a complete history of Chagas disease in Argentina, see Zabala (2010), to whom I am grateful for details about the process in question.
4. Compulsory military service was abolished in Argentina in that year. It was the only formal occasion when the health of the population was systematically controlled. Although it was restricted exclusively to young men, it covered the entire population.
5. The figures are from the National Institute of Statistics and Census of Argentina (INDEC) for 2001 (www.indec.mecon.ar). Extreme data are taken in order to show the contrast, while taking into account, of course, that life expectancy for women is slightly higher than for men.
6. His pioneering work of 1975 on the scientific field was followed, in his last years, by a text on the "social uses of science" (1997) and his last course at the Collège de France, published under the title "Science de la science et réflexivité" in 2001.
7. We can see, in this trace, elements of the functionalist sociology of science, of which, in the figure of Robert Merton, Bourdieu was once a steadfast defender. See his little-known text "Animadversiones in Mertonem" (1990).
8. Although, more straightfworwardly and directed at public policy, Oszlak and O'Donnell (1995 [1981], p. 111) framed similar issues in the same period. For these authors, the aim was to identify the "emergence of an issue": who problematizes a question and how; who—and how and when—manages to turn it into an issue; with what resources and alliances; with what opposition; what is the initial definition of the issue.
9. In the mid-1940s, Chagas disease was defined as a "national disease", and the creation of a national program was the result of that stance. See Zabala (2010) and Kreimer and Zabala (2006).
10. Strangely, this last example is not pure fiction; these were a presidential candidate's opening statements in Argentina a few years ago.
11. See the purification schema, and a deeper explanation of this process at the end of Chapter 3.

5 Rowing Against the Tide? New Research Fields in Peripheral Contexts

Molecular Biology From the Pioneers to the New Genomics (1957–2017)

Introduction

In 1957, the first molecular biology laboratory in Argentina was established in the National Institute of Microbiology (*Instituto Nacional de Microbiología Dr. Carlos Malbrán*). It was headed by Cesar Milstein, a young chemist who won the Nobel Prize in 1983 for his work on monoclonal antibodies at the Medical Research Council (MRC) Laboratory in Cambridge.

However, five years later, following a decision by the Ministry of Public Health, the laboratory was dismantled, and most of its researchers emigrated or moved to other fields. There is an important question to pose here: Why was a molecular biology laboratory set up in the late 1950s, in a peripheral context, at a moment that could be described as being "early" in the development of molecular biology as a scientific discipline?[1]

This question is the crossroad of several issues: the disciplinary level of analysis, as we discussed in Chapter 2; the center–periphery relations, discussed in Chapters 2 and 3; and the utility of knowledge, discussed in Chapters 4 and 5. Consequently, molecular biology in Argentina is a fascinating example of the development of a new knowledge space in a peripheral context.

The development of molecular biology in Argentina happened quite early, with the discipline just taking off internationally: indeed, there were very few groups around the world working on topics that could yet be called "molecular biology". For the scientists involved, it was not yet clear that the techniques issuing from molecular biology were laying the foundations of a new discipline. On the contrary, many accepted the new research techniques without realizing that the investigational logic of the life sciences was being called into question. This point has its correlate on the international stage, where the term "molecular biology" was not perceived as denoting any specifically defined field of knowledge.[2]

Scientific practices are generally formulated at the micro level of workspaces, including "laboratory cultures and local traditions rather than national context [. . .] institutional settings, schools and informal net-works" (Gaudillière 1993). They are also shaped by local traditions, as they form networks involving different generations, collective spaces, and genuine "transepistemic arenas" (Knorr 1982) where the interplay between local actors is particularly important.

The setting up of a molecular biology laboratory in Argentina can be viewed as a precedent for the construction of a new field of knowledge and social practice. Its history involves not only "disciplinary innova-tion" but also factors that shaped the development of molecular biology in Argentina and its relationship with the international scientific com-munity. A particular feature of "peripheral" science we have described as "subordinated integration", a concept we have already discussed in Chapter 2.

It is usual, when periodizing the history of science in Argentina, to establish a relationship between military intervention, scientific emi-gration, and institutional instability. The irruption of politics into the realm of science has been a defining factor in Argentina, as it has in other Latin American countries (and certainly other developing countries as well), and an understanding of the years 1943, 1966 (the year of the Onganía dictatorship's violent intervention in the University of Buenos Aires, known as the "Night of the Long Sticks"), and 1976 is vital to our understanding of the dynamics of science in this country. Even so, the history of science cannot be seen entirely in terms of politico-military interventions. Such an approach tries to account for the professionaliza-tion and institutionalization of scientific practice and the emergence and consolidation of a scientific community. It does not distinguish between different disciplines and is based on the (partially true) supposition that most (human and material) resources are concentrated on the university space. However, not every politico-military intervention had equivalent effects on every discipline. Nor were these interventions processed identi-cally by the various institutions and actors involved. My approach here rests on understanding the specific schisms and continuities of the groups that provided molecular biology in Argentina with its impetus. I think it is necessary to establish elements arising within both the socioinstitu-tional context and local scientific practice. Thus, while "external" inter-ventions necessarily emphasize ruptures, a more comprehensive history has to explain both ruptures and continuities.

To understand the changes in molecular biology's social and cogni-tive organization, Terry Shinn's (2000, pp. 60–61) categories for what he calls scientific and technological "research regimes" are useful. Beginning from the identification of social and intellectual mechanisms, he posits three different regimes: disciplinary, transitional and transversal. Let us look briefly at these categories.

Disciplinary regimes normally appear in easily identifiable, stable institutions. Their roots lie deep in laboratories, university departments, specialist journals, national and international conferences, and official reward systems. Indicators make it easy to detect and analyze precise career models and differentiated categories of scientific production.

In transitional regimes, intellectual, technical and professional opportunities often arise on the periphery of the classical disciplinary fields. In this case, setting up or doing a degree course demands that practitioners, temporarily crossing the frontiers of their own discipline, seek out techniques, data, concepts, and cooperation from colleagues in adjoining fields. Much of the time, the search for cognitive, material or supplementary human resources involves two or three disciplines. Movements are inscribed within an oscillatory model of comings and goings. In transitional regimes, practitioners' main center of identity and action is still bound to the disciplines, while individuals cross disciplines.

In transversal regimes, knowledge production is different. Practitioners' degree of freedom and field of action is greater than in transitional regimes. Yet, at the same time, it is difficult to record data about their career trajectories. Such difficulty partially explains the dearth of studies on this dynamic of research. Thus, the practitioners of transversal regimes identify more with off-the-cuff projects than with the disciplines and institutions they inhabit. Practitioners' range of action is wide and enables free movement in social and material space.

These three regimes, says Shinn (2000, p. 65), may be viewed as interdependent and enriched by reciprocal interplay, and it is certainly possible to find common ground. They are all founded on one form of division of intellectual, technical and social labor. The boundaries between scientific research regimes and other sectors are important, for they enable the researcher to define objectives and rivalries and to pull through any onslaughts or hard times. These boundaries act as both a corporate defense system and a mechanism that gives access to privileges and social ascent.

We must make one or two special remarks about disciplinary regimes that will be of special significance in a case presented in the third part of this chapter. The wide range of studies available has made it particularly important to specifically focus on the institutions where this kind of regime occurs: academic institutions where social actors are deeply imbued with administrative regulations and practices have clearly prescribed institutional foundations. The legitimation of knowledge is therefore essentially endogenous: first, the institution itself that gives its practitioners a "credential" that acknowledges them as legitimate researchers; and, second, the legitimation of the community of practitioners who judge the techniques and concepts of the practices of those who are working within a specific discipline. The community of practitioners always forms as a local and international reference organized by specific institutions (international associations). These in turn regulate the operation of a disciplinary framework and are usually the space where struggles over the field's direction and boundaries take place.

Therefore, in line with the dynamic of research groups in Argentina, we have laid out five periods:[3]

1. Pioneers (from 1957 to 1962): The creation and development of the first laboratories at the Malbrán Institute.
2. "Vacuum" (from 1962 to the early 1970s): Most researchers who had been part of the earlier experience either went to work abroad or dropped their research into molecular biology.
3. Consolidation and establishment of a "transition" regime (from the early 1970s to 1982): The creation of the first institute dedicated entirely to research into molecular biology.
4. The crossover to a "transversal" regime (from the early 1980s to the mid-1990s).
5. The new shift: The era of the genetics.

The structure of the chapter is as follows: In the first part, I shall briefly present the main features of the context and the institutions. The second part shows the development of molecular biology in the world between 1940 and 1950 and the research traditions that emerged. The following sections focus on these five periods and, finally, we finish with a general conclusion.

Context and Institutions

Political and Cultural Modernization in Argentina

Beginning in the mid-1950s, Argentina underwent a process of modernization and institution-building. This was accompanied by a convergence between research and science policy. Generally speaking, this innovation was characterized by a steady transformation of older institutions and the creation of several new ones. Among the newer institutions were four institutes that, along with the national universities, formed the backbone of the Argentinean scientific and technological complex.

The institutionalization of science policy making was influenced by several concurrent features, including (a) a post-Peronist revival in the universities; (b) the wider application of development models emphasizing the role of science in economic planning; (c) the influence of the Economic Commission for Latin America and the Caribbean (ECLAC) and its theories of economic development by import substitution; (d) the institutionalization of science and technology policy overseas, especially in France; and (e) an increasingly active set of international agencies, notably UNESCO and the OAS.

Among the new institutions in Argentina, the most important was the National Scientific and Technical Research Council (CONICET), set up in 1958 to "promote, coordinate and steer research in the fields of the pure and applied sciences and technologies" (Decree-Law N° 1291/58).[4]

The prevailing idea was that a *system* of science and technology would integrate disparate elements and take "the mobilization of scientific and technological knowledge for development" as its overall goal. In Argentina, this process was led by Bernardo Houssay, CONICET's founder and first president. Houssay, a physiologist who won the Nobel Prize in 1947, had a good reputation within the local scientific community and with the political leadership. CONICET never wielded the function of "policy design and planning" assigned at its inception. However, it did make extremely vigorous efforts at promotion, through the award of grants and scholarships, subsidies for research groups, and the creation of a "status system" (similar to that of the CNRS), which guaranteed career stability. Eventually, CONICET set up institutes of its own, but this practice was marginal until Houssay's death in 1971 (National Secretariat of Science and Technology, SECyT, 1989).

After Peron's fall, the year 1957 saw the beginnings of reform at the University of Buenos Aires. This was a time when many researchers who had been sidelined (or expelled outright during the Peronist period) returned home, including Houssay and Leloir.[5] University reform involved modifying teaching methods and introducing practical work at the expense of lecturing. Syllabi were made more flexible, and faculties were replenished with new full-time posts. Teaching reforms were accompanied by new research institutes, such as the Calculation Institute in the School of Exact Sciences. New schools in pharmacy and biochemistry were also created, as were new degrees in psychology and sociology (Halperín Donghi 1962; Sigal 1991).

Although the state allotted resources and regulated institutional activities, there was no structured or clear-sighted plan in this development. Still, the idea that science and universities could play a central role in socioeconomic improvement gathered force. It was presumed that, when research reached "critical mass", it would produce "lift-off", with direct benefits for Argentina. The ideologues of this movement were the leaders of the "academic and scientific communities" (Vessuri 1994a, 1996a). Institutions such as CONICET and the universities were structured according to a view that held that scientific activity was superior to "second-order" activities, such as establishing links with the secondary industrial sector.

The creation of an institutional context for experimental research in the biomedical disciplines was inevitably linked to this tradition. In the process, it called upon external recognition. This was essential to research, but inevitably involved a tension between predominant trends in "international science", local socioinstitutional conditions, and the needs of local users.[6]

During the late 1950s, there were two differing positions regarding the role of scientists.[7] On one side, Houssay represented a "traditional" view of science, performed with cheap instruments often made by researchers themselves (the "bricoleur scientist"). For Houssay, resources were to be allocated on the basis of individual requests and always in "reasonable" amounts. On the other side, there were researchers linked to the UBA's Exact and Natural Sciences School, led by its dean, Rolando García, who took a more modern view: They had seen the changes generated by "big science" and propounded the need for greater resources organized around planned activities with long-term goals. For Houssay, this position amounted to "setting up a Soviet system".[8] Broadly speaking, it was Houssay's position—described as "scientificist" by his adversaries—that eventually won the day. Houssay believed that,

> while he had always been in favor of research carried out in universi-
> ties, they were at the time arenas for political debate where decisions
> were taken by vote and usually swayed by trade union pressure and
> shady political dealings, any funds allotted to researchers by insti-
> tutions such as the CONICET being swallowed up by university
> budgets, and universities generally making the most of this foreign
> backing to stop supporting science with their respective funds.
> (Cereijido 1990, pp. 95–96)

Houssay recalls that "extramural institutes were oases where, when times were hard, many scientists repopulating the universities [he himself was a case in point] took refuge" (Cereijido 1990, p. 96).[9]

The Institutional Space of the Malbrán National Institute of Microbiology

Within this broad process of institutional renovation, the Carlos Mal-brán Institute, created in 1916, was reorganized in 1957 as the National Institute of Microbiology.[10] At its inception, the institute aimed, first, to produce sera and vaccines, with a view to making Argentina self-sufficient in these products; second, to keep tabs on glandular extracts sold to the pharmaceutical market; and third, to monitor, control and study epidemics and diseases (including the drawing of an epidemiologi-cal map). Such "technical" tasks were combined with research, which began to appear in a journal published by the institute.[11]

In the early days, there was a gulf between "pure" research and the applied development of diagnoses and serum and vaccine production. This gulf began to widen as the institute's different specializations became more professional. It is interesting to compare the Malbrán Institute with the Oswaldo Cruz Institute in Rio de Janeiro, Brazil, since both were modeled on the Pasteur Institute in Paris. As Nancy Stepan has observed,

Cruz "expressed dissatisfaction with the Institute's restricted function as a supplier of vaccines and serums. From his training at the Pasteur Institute, Cruz acquired a working knowledge of the way one of the most scientific institutions in the world was organized" (Stepan 1976).[12] However, the development of the institution in Brazil should not be seen as a mere copy; the transfer process turned out to be far more complex. Significantly, in Brazil there was no trained cadre of bacteriologists to serve the dual function of basic and applied research.

The "Pasteur" was special in the multiplicity of functions it embraced— the diversification of its basic laboratories, the production of biological material, epidemiological control and vigilance (*veille*), a hospital and a space for training researchers in conjunction with the University of Paris (Institut Pasteur 1987). It was a unique institution, beginning life after a public auction and emerging as a private institution "orientated toward the public sector". Twentieth-century public health policy could not turn its back on the institute's privileged status. The institute gradually became a legitimizing space for professional identification, both in France and abroad.[13]

In a similar way, bacteriology in Argentina became professionalized between 1924 and 1944, when Alfredo Sordelli was director of the Bacteriological Institute. Sordelli was a highly regarded pioneer in research, and, with Houssay, a driving force in the transition from "traditional" physiology and medicine towards the experimental approach emerging in the School of Medicine in Buenos Aires and some of the city's hospitals. In this period, the institute's structure was defined. The production of sera and vaccines was combined with private practice and pure research. With the exception of a handful of department heads in the institute's upper echelons, staff did not devote themselves to full-time research. The professionalization promoted by Sordelli ensured that the mechanisms to encourage research were based on conducting an open competition[14] (entrance examination), on sending scholarship holders abroad and on inviting leading figures to shape the new departments.[15] Eventually, the organization was based upon a division of work into different autonomous departments, in which the establishment of priorities gave each director great autonomy.

During this period, attempts to promote full-time research exposed tensions between competing institutional models. On the one hand, the outbreak of epidemics required the institute to be a center for public health. On the other hand, the research model proposed by Rudolf Kraus, which bore fruit under Sordelli, led to the development of an incipient research community and to important relationships with research centers abroad.[16]

With this combined function, the Malbrán Institute began to take shape between 1956 and 1962.[17] In 1956, Ignacio Pirosky was appointed interim director by Dr. Francisco Martínez, minister for Social Welfare and Public Health, with the explicit aim of reorganizing the institute.

Pirosky felt that the "first issue to address was the economic issue. It was essential if we were to stay in contact with the major scientific centers around the world". Furthermore, Pirosky backed the director of public health in securing the "necessary elements for the manufacture of top-quality sera and vaccines and the strict monitoring of the efficacy of certain biological medication. But principally the Institute *had to* become a center for technical training and scientific research". (Pirosky 1986). A year after Pirosky became director, the old Bacteriological Institute became the National Institute of Microbiology. The new institution's goals were fivefold—to perform scientific research in the various microbiological disciplines in the spheres of both pure research and its application to public health, to manufacture serums and vaccines; to carry out microbiological diagnoses, to study the etiology of endemic and epidemic diseases in the country, to promote active exchange with the major microbiological centers around the world, and to promote a system of grants and scholarships.[18] Within this scheme, the development of research, as well as the coordination of the departments, was the exclusive responsibility of the director.[19] A modernization process was begun, giving priority to the development of basic research in serum and vaccine production.

The first experiment in molecular biology research in Argentina became operational on the following principles:

- The institute purchased state-of-the-art equipment to encourage research.[20]
- An open examination was held to generate full-time technical-scientific staff; from this, ninety new researchers were taken on.
- Grants and scholarships were promoted for overseas training (approximately thirteen researchers left the country).[21]
- New research spaces were created for these young researchers so that they would be able to reinsert themselves upon their return to Argentina (this was especially important for those who worked in the molecular biology department, which would also contain the bacterial genetics department).

In one sense, these principles of modernization led to cognitive changes inside the institution by substantially modifying the institution's structure. These efforts were influenced by events in Argentina that opened fresh opportunities for tackling cognitive issues in different fields.

Emerging Research in Molecular Biology

Molecular Biology in the Late 1950s

To put these laboratories into context, it is worth summarizing the state of the field at the time. Although the term "molecular biology" was in use in the 1930s, it did not become widespread until the end of the 1950s, with the appearance of the *Journal of Molecular Biology*.[22] During the 1950s, the expression did not yet signify a given field, but rather an innovation designating a new *hybrid* that could not as yet be identified, even by the scientists involved. Francis Crick, who proposed in 1953—with James Watson—the double-helix model for the understanding of the DNA structure, stated that

> I myself was forced to call myself a molecular biologist because when inquiring clergymen asked me what I did, I got tired of explaining that I was a mixture of crystallographer, biophysicist, biochemist, and geneticist, an explanation which in any case they found too hard to grasp.
>
> (Crick 1965, p. 183, quoted by Stent 1968, pp. 390–395)

Only when the field began to establish itself were its limits described with any clarity. With the publication of *Phage and the Origins of Molecular Biology* in 1966, practitioners saw that a new discipline had emerged and began to trace its development.[23] Their history began with a succession of disciplines, converging during the 1930s, that gradually brought about the acceptance of a new level of analysis (the molecular level) as the basis for understanding reproduction in living organisms.

After the Second World War, new science policy initiatives facilitated scholarly travel and the establishment of international networks. These in turn produced the kind of transnational context that made possible the key consolidating discoveries of molecular biology—namely, the structure of the double helix and the RNA messenger. If we take as a guide the predominant tendencies described by the German biologist, Gunther Stent, we can identify three separate fields at the end of the 1950s:

1. *The Structural*: The study of the architecture of biological molecules, taken up by three groups: the first, headed by Linus Pauling at Caltech; the second, by Max Perutz (a disciple of Bernal) at the Cavendish Laboratory in Cambridge; and the third, led by Rosalind Franklin and Maurice Wilkins, at King's College, London.
2. *The Biochemical*: The study of the interaction of biological molecules in cellular metabolism and heredity, developed by the French group consisting of Lwoff, Jacob, Monod and Wollman at the Pasteur Institute in Paris.
3. *The Informational*: The study of the ways in which information is conveyed from one generation of organisms to another and the translation of information into biological molecules, developed by the so-called "Phage Group", led by Delbrück, Luria, and Hershey (Stent 1968, pp. 390–395)

According to Stent, the period between 1953 and 1963 saw a "dogmatic phase", dominated by James Watson and Francis Crick. (Stent 1968, pp. 393–394). The theoretical importance of their discoveries stemmed from the fact that the search for the double helix was based on the introduction of genetic reasoning into structure determination. This later translated into the central dogma of molecular biology. In this period, there were other important events in the consolidation of molecular biology, such as Lwoff's discovery of operon and the RNA messenger. Likewise, the English school of x-ray crystallography established the molecular structure of hemoglobin (Perutz) and myoglobin (Kendrew), while in the biochemistry laboratory at Cambridge, Sanger established the first complete sequence of amino acids for a protein (insulin). All these (Nobel Prize–winning) discoveries gradually shifted attention from proteins to nucleic acids and DNA (Kendrew 1967).

Molecular Biology Laboratories

Argentina's first molecular biology laboratories opened in 1957, when Dr. Ignacio Pirosky was appointed director of the Malbrán Institute.[24] This event could be viewed as an example of "institutional transfer", inspired by the Pasteur Institute, where Pirosky had worked under Lwoff in the 1930s. However, the Malbrán Institute found it extremely difficult to make headway in basic research due to both political and professional limitations. Behind the call to provide staff positions, two institutional models were in conflict: the traditional model of the Malbrán as an institute for services, and the Pirosky model, which saw the institute as devoted not only to applied research (including epidemics control and vaccines production) but also to basic research.

The first laboratories involved a shift towards a new way of thinking. The most visible effect of the recruitment process was the promotion of new, full-time posts. However, this generated occupational conflict among the representatives of the "old guard" (staunch supporters of tradition within the institute) and the new generation entering the institute in Pirosky's recruitment.[25] After 1957, new departments were tailor-made for new staff and researchers returning from postdoctoral sojourns abroad. A department of bacterial genetics was added in 1957 and—upon Milstein's return from England in 1961—a molecular biology department. Milstein was to head this department until 1962. Molecular biology subsumed the bacterial genetics department, as well as a group of previously independent researchers.

These creations did not rupture the institute's structure into autonomous departments. Rather, they emerged gradually within an organization that was based not upon any coherent design, but upon whatever orientation the director of the day decided. The arrival of molecular biology, with its innovative set of devices and techniques, brought with it a cognitive rupture. However, this rupture was not accompanied by a consequent change in organizational form, and to a large extent, traditional institutional settings prevailed. Of the two new departments, the first was in bacterial genetics. This comprised the biologists Rosa Nagel and Juan Puig.[26] A year later came the geneticist Dora Antón, from the School of Exact Sciences at the University of Buenos Aires, where she had held a post in the Department of Genetics under the agriculturist, Juan Valencia.[27] Until then, the field of Argentinean genetics was dominated by agriculturists interested in hereditary features in crop varieties. With the mid-1950s came the first developments in microbial genetics. The creation of the bacterial genetics department at the Malbrán Institute brought about significant thematic and conceptual innovation.[28] Investigations at the molecular level, using virus systems (phages), established DNA as the "transforming principle".[29] A line of research centered on the study of prophages and the features of RNA messengers, in which the "French school", represented by Lwoff, Monod, Jacob, and Wollman, played a dominant role.

Elie Wollman's participation was crucial to the new laboratory. Wollman visited the old Institute of Bacteriology in 1929, when his parents were invited to establish a link between the Malbrán and Pasteur Institutes (Personal letter from Elie Wollman, June 2000). Collaboration followed, and Pirosky was invited to the Pasteur Institute to work under Lwoff in 1936 and 1937. Pirosky's presence responded to an explicit strategy of incipient "internationalization", supported by the Rockefeller Foundation, whose program director exerted a great influence over the young Jacques Monod.[30]

According to Wollman,

> when Pirosky became director of the Malbrán Institute, he went to the Pasteur Institute, and told Lwoff about his plans for innovation and, in particular, for the recruitment of a small number of young Argentinean scientists who wished to turn to bacterial genetics. Lwoff, who knew about my long-standing links with Latin America, urged me to accept.
>
> (Personal letter from Elie Wollman, June 2000)

As a result, Wollman went to Buenos Aires in August 1958 and enlisted Juan Puig, Rosa Ángel, Pablo Bozzini, and Dora Antón. The group set up in the "Pasteur Pavilion" of the Buenos Aires Institute. Wollman brought the biological material and devoted himself to the laboratory while lecturing in the Faculties of Science and Agronomy. Puig, Nagel, and Antón presented their doctorates the following year. Unlike the bacterial genetics laboratory, the creation of the Molecular Biology Division, headed by Cesar Milstein, was linked to the so-called "structural", or British, tradition in molecular biology. Milstein worked with a medical doctor, Andrés Stoppani, in the School of Medicine, where Stoppani headed the Institute of Biological Chemistry. In 1957, he entered the Malbrán Institute, where Pirosky gave him leave to do a doctorate at Cambridge. At Cambridge, molecular biology revolved around two groups—the first, led by Perutz and Kendrew at the Cavendish, specialized in x-ray diffraction of molecular structures of proteins, while the second, headed by Frederick Sanger at the Biochemistry Department, specialized in new chemical techniques and amino acid sequences. Milstein went to Sanger's group, where he worked on enzyme activation and amino acid sequences in active enzyme centers. The experience gave him a knowledge of protein chemistry and kinetic enzyme analysis (Interview with César Milstein, Cambridge 10 January 1999).

When Milstein returned to the Malbrán Institute, Pirosky appointed him director of the Molecular Biology Division. This consisted of Milstein, Noel Zwaig (Rosa Nagel's husband, who worked in the division founded by Wollman), Marta Pigretti, Celia Milstein, Manuel Brenman, Nazario Mahafud (a scholarship holder), and Horacio Farach, Teodoro Celis, Inda Issaly, and Abel Issaly (all writing their PhD dissertations). Under Milstein's management, bacterial genetics merged with this division; however, in practice, the laboratory was a great deal more independently than other groups run by Milstein.

Its first line of research was directed by Milstein, who continued what he had done at Cambridge and took further the development of techniques for the study of sequencing and marking the active centers in phosphoglucomutase, phosphoglyceromutase, and alkaline *E. coli.* A second line of research was directed by Dr. Manuel Brenman, who specialized in nuclear magnetic resonance. His group began in late 1961, after Brenman returned from studies in nuclear magnetic resonance in the Chemistry Department at MIT.

Cognitive Liberalization and International Referents

Wollman's leadership and Milstein's return from Cambridge were key factors in the emergence of the new discipline in Argentina. Lwoff's group at the Pasteur Institute, together with Perutz's at Cambridge and that at Cold Spring Harbor in the United States, were among the few devoted to the new discipline. Thus, of the three major international groups, molecular biology in Argentina had close ties with two. The formation of the new laboratories was also linked with one of the privileged forms of innovation in the Argentinean scientific community. A form of integration was achieved by subordinating Argentinean science to mainstream international science.[31] In this peripheral context, there were two ways of setting new agendas—by bringing in scientists from abroad, and by hiring researchers who had completed their studies overseas.[32] Both strategies were employed.

Pirosky sent the institute's scholarship holders to centers specializing in relevant areas. In every case, he kept up-to-date with their progress, their equipment, and their research so that they could continue work they had been doing abroad. The strategy deployed by Pirosky in sustaining a network with centers of scientific excellence working in the field of molecular biology was tied in with a strategy to use foreign recognition to improve local standing. In fact, much research was done under Pirosky, who succeeded in finding funds to pay full-time salaries and purchased scarce equipment.[33] These resources came mainly from the Ministry of Public Health, thanks to support from the minister of the day and, to a lesser extent, from the recently created CONICET, on the board of which Pirosky was a member.

Although he did not align himself politically with the hegemonic group in CONICET (Bernardo Houssay and his disciples), but rather with the "innovative group" headed by Rolando García (dean of the Exact and Natural Sciences School in the University of Buenos Aires), Pirosky's presence did institutionalize the groups at the Malbrán Institute as a legitimate space recognized by the Argentinean scientific community at large. The existence of an established scientific tradition in biomedical research became the cornerstone of conceptual innovation in molecular biology. Even if scientists did not realize that molecular biology actually represented an innovation, they tended to favor new research techniques that broadened without disrupting "traditional" biology. Moreover, during the 1950s, biochemistry was dominant within the local community, and this was a limiting factor.

The last and most important point concerns the relationship between the Malbrán laboratories and the local scientific community. The movement, spurred by Pirosky, was inscribed within a tradition represented by Alfredo Sordelli, as well as within a network of modernizers, who went beyond this tradition. Sordelli was identified with a generation that institutionalized experimental research through methodological rigor and international networks.

This institutionalization began in the early twentieth century, was temporarily challenged during the Peronist period, and then re-emerged in 1956. In this sense, close relations with the Cori couple and Severo Ochoa in the United States, with the Pasteur Institute in Paris, and with the Rockefeller Foundation, all indicated a modern way of doing science that had taken root in Argentina. However, by the end of the 1950s, this model was looking out-of-date, at least when compared with two transformations on the international stage—the emergence of "big science" and the impressive development of the biomedical sciences, which made intensive use of new techniques and equipment.

The Laboratories' Demise

The Malbrán's molecular biology laboratories were dismantled in 1962, when Pirosky was removed as director by the newly appointed minister of public health. This led to a breakdown in a process that had come of age in the years between 1957 and 1962. A number of measures completed the breakdown: full-time posts were limited and most of the staff were dismissed.[34] In the circumstances, Pirosky's institutional project, linking public health objectives with the development of research, found itself in competition with the type of specialization defended by "the old guard". This model was backed by the new authorities at the Ministry of Social Welfare and Public Health, as well as by the School of Medicine at the University of Buenos Aires.

In reviewing the Malbrán Institute, they reported that

> the activity attributed to this body has not been carried out, where aspects directly relating to the problems of public health are concerned . . . that, therefore, supplies of certain essential products have in some cases been outstripped by needs [. . .] and that the necessary measures have not been taken to solve the shortcomings in production processes, while other measures did not fit with the needs of the Institute.[35]

This apparent failure to serve its appointed function led to the dismissal of twelve researchers, six of whom belonged to the molecular biology department. This, in turn, led to the resignation of another thirteen researchers, six of whom were in the same department. Thus, the first molecular biology laboratories in Argentina were effectively disbanded.[36]

Following their departure, these pioneering researchers continued their careers abroad. The development of new techniques and the modification of practices involved in the introduction of the new science of molecular biology were virtually lost to Argentina, as the researchers dismissed from the Malbrán Institute went abroad: Milstein returned to Cambridge; Nagel, after a short period as professor of genetics at the University's School of Exact Sciences, went to the United States to work with Luria; and Puig joined a CNRS Institute in Marseilles. Luria and Puig returned to Latin America at the end of the 1970s, but to Venezuela, not to Argentina, where they could not return because of the military dictatorship installed in 1976. Despite its pioneering prospects, molecular biology was lost to Argentina for more than a decade. It was impossible to nourish the discipline in the university schools of exact sciences or medicine, or in laboratories belonging to CONICET.

The "Vacuum"

Thus, research into molecular biology in Argentina reached a decade-long *impasse* (or "vacuum") lasting until the early 1970s, when the first molecular biology groups, particularly at the Campomar Foundation, began to appear (see Figure 5.1). The first molecular biology and genetic engineering laboratories were organized and consolidated between 1971/1972 and the mid-1980s. This was the result of a slow process of generational renewal taking in the "first generation" of Leloir's disciples, then the disciples of that generation.

Of course, the idea of a vacuum is relative. It refers particularly to the disappearance of the pioneering groups and the lack of any *systematic* research in the field over these years. During this decade-long impasse only a few isolated, incipient groups tackling molecular biology–related problems survived: the Molecular Genetics Laboratory in the Biology Department of the Faculty of Exact and Natural Sciences of the UBA headed by José L. Reissig and the molecular biology group headed by Gabriel Favelukes at the then Faculty of Chemistry and Pharmacy of the National University of La Plata.

The first of these laboratories operated between 1960 and 1966, when intervention at the UBA caused Reissig's removal and subsequent departure abroad, the dispersal of the research group, and the disappearance of the Molecular Genetics Laboratory.[37]

The second came about early in the 1960s, when Favelukes returned to the National University of La Plata from the United States, where he had specialized in early in vitro protein synthesis. Although he formed a group that, to this day, has been working on various lines of research, the shortage of initial funding and lack of a conducive institutional space prevented rapid integration and subsequent consolidation. Even so, in the early 1970s the Favelukes group succeeded in organizing the first pioneering molecular biology conference in Argentina.[38]

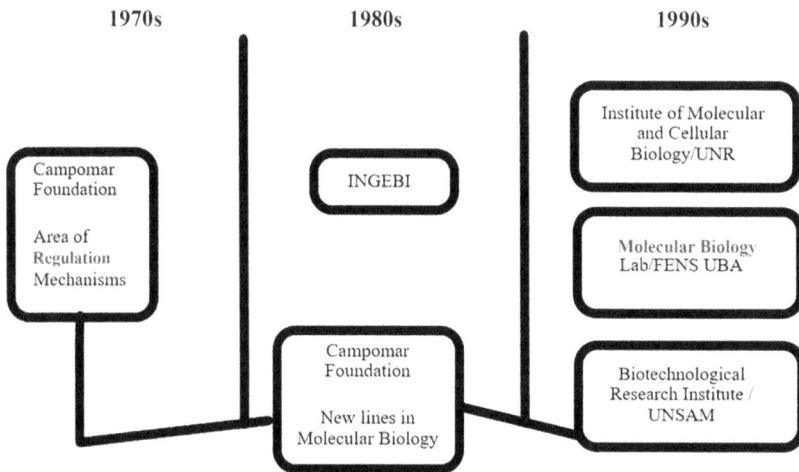

Figure 5.1 Molecular Biology Groups from the Campomar Foundation (1970–1990)

To institutional weakness, one must add the sociocognitive restrictions involved in disciplinary schisms. As we have already pointed out, the crystallization of biochemistry as a "central discipline" in the biomedical field blinkered people to the autonomization of a new international discipline. Molecular biology appeared to the actors of the day (particularly Leloir and some of his students) as a set of novel (and useful) techniques for solving problems regarding metabolism and cellular transformation from the standpoint of biochemistry.

Molecular biology effectively became an autonomous international discipline during this period, with research centers specially dedicated to these questions and new ways of associating, particularly in the form of new international cooperation and collaboration networks. It was in the context of molecular biology's international consolidation framed by a local vacuum in Argentina that an incipient generational renewal began to take place at the heart of the biomedical tradition in this country. Although many of these researchers left to specialize in biochemistry or other traditional fields, the discipline's rapid expansion meant that these young scholarship holders steeped themselves in the conceptual framework of molecular biology.

Consolidation: The Development of a "Transition Regime"

According to our periodization, the first systematic approaches to molecular biology date back to the 1970s. This was when the first research group was formed. Led by a disciple of Luis Leloir, it was devoted entirely to research and not only used new techniques but also formulated new problems that were different from traditional biochemical concerns.

The group was run by Israel Algranati, who returned to Campomar in 1969 after finishing his postdoctoral studies with the Spaniard Severo Ochoa.[39] Significantly, Algranati was being trained abroad just as the field's conceptual limits were being laid down (under what Francis Crick defined as the *Central Dogma of Molecular Biology*). Here, the principle of *colinearity* is formulated in that a dual univocal correspondence is set up between the expression of a gene and the synthesis of a given protein. Thus, from the early 1970s, research concerned with the relationship between nucleic acids and protein synthesis emerged. This research would later address the deciphering of the genetic code. Among the researchers involved were Severo Ochoa (School of Medicine, University of New York), Marshall Niremberg (National Institutes of health, Bathesda), Har Khorana (University of Wisconsin), and Robert Holley (Cornell University). In 1968, the last three researchers received the Nobel Prize for Medicine for interpreting the genetic code and its workings in the synthesis of proteins.

The fact that Algranati went to work with Severo Ochoa was due, on the one hand, to his having worked on protein synthesis–related topics since joining the Campomar Foundation, especially on the role ribosomes play in the synthesis or degradation (metabolism) of carbohydrates. It was also due to the long-standing relationship between Leloir and Ochoa. Once back again at Campomar, Algranati put together a group along his original research lines, though altering the frame of interpretation for the biological function. Algranati consequently adopted principles of molecular biology (*all the information regulating protein synthesis is contained in the genes, and this synthesis depends on how these are expressed*) in order to understand how cells went from states of rest to being metabolically active. Molecular biology enabled him to inquire into new problems that were nevertheless clearly defined and complemented the field of structured problems from a traditional biochemical way of thinking: the glycolization of proteins and the synthesis and decomposition of sugars and oligosaccharides. Molecular biology was thus incorporated into the biochemical paradigm as a topical innovation rather than an autonomous field of knowledge. Indeed, it was comprehended mainly as a set of (auxiliary) techniques that broadened the biochemical paradigm's field of studies. Other groups soon joined this first group, forming an area of "regulation mechanisms".

In all these lines of research, a dual relationship is to be observed between molecular biology and biochemistry, and indeed, early on these groups stuck closely to the latter. This adjustment of a new field to the limits imposed by a traditional field is what enabled the first group of the biomedical "central nucleus" to delve into the new discipline in more depth.

The consolidation of an autonomous molecular biology stems from a paradox: The process began at the same time Leloir received the Nobel Prize for Chemistry in 1970. This permanently places Leloir and his institute, the Campomar Foundation, at center-stage in the biomedical tradition initiated by Houssay. Yet the reaction of Leloir and his closest collaborators to the new field's emergence was never favorable. At first, the organization of the newly formed lines of molecular biology research reflected earlier modes of organization. This was a period of transition that relegated the new discipline to the role of "auxiliary techniques or topical innovations", rather than a relatively autonomous field.

Young researchers joining the institute in those days, mostly in the area of regulation mechanisms and especially Héctor Torres's group, conceived of molecular biology as involving a new conceptual framework requiring a different level of knowledge from traditional biochemical frameworks. One of the most significant leaders of the "new generations", a biologist entirely "socialized" in the new paradigm, held that "this set of methodologies is no mere technological recipe book, as for a long time some believed, but a new way of thinking about and solving biological problems".[40] This conception of molecular biology would take shape in parallel with these young researchers doing their postdoctoral studies abroad in the second half of the 1970s in a context of profound changes in the discipline associated with the new genetic engineering and biotechnology techniques (Sasson 1988).

Consolidation of the discipline entered its "mature" phase in 1983 with the setting up of the first institute dedicated entirely to molecular biology. This was the Institute of Genetic Engineering and Molecular Biology under Héctor Torres. The new institute grew up out of the Torres group's separation from Campomar. Several researchers who had worked with him there went on to form part of the new institute's staff: Mirtha Flawiá, María Téllez Iñon, Alberto Kornblihtt (returned from England), Gerardo Glikin, Luis Molina y Vedia, and Alejandro Mantaberry (who worked at Carminatti's glycoprotein laboratory). Over the years new researchers were taken on, some of whom did various studies at molecular biology centers abroad: Mariano Levin (INSERM/Pasteur Institute, Paris), Alejandro Paladini, Luis Jiménez de Asua (Molecular Biology Laboratory, Salk Institute for Biological Studies), Marcelo Rubinstein, Adolfo Iribarren (European Molecular Biology Laboratory and the Research Institute of Molecular Biology, Italy), Rita Ulloa Paveto (Massachusetts University), María Rossi, Liliana Finocchiaro, and Héctor Chuluyan.

Meanwhile at the Campomar Foundation in the 1980s, a growing number of lines of research were set up incorporating new topics or techniques from molecular biology. This gave molecular biology ever greater penetration:[41] Oscar Burrone (who had started out life with Algranati) formed his own group in molecular immunology and virology, Alberto Frasch in molecular studies of *T. cruzi*, and Roberto Staneloni in the regulation of genetic expression in higher plants, for example.

New areas of research whose cognitive logic went beyond the mere addition of "new lines" arose from foreign study trips made by the new generations of researchers. This group was responsible for the manifestation of molecular biology as an autonomous field of knowledge locally. They brought with them new problems (genetic expression, the transformation of genetic material) and research techniques (sequencing, molecular markers) and displaced the focus of attention to the genetic structure and transformation of organisms. The process culminated with the formation of new laboratories and institutes directed by researchers trained at Campomar or the INGEBI between the late 1970s and early 1980s. This was the case with the Physiology and Molecular Biology Laboratory led by Kornblihtt at the UBA's Faculty of Exact Sciences, the National University of San Martín's Institute for Biotechnological Research with Frasch (the author of one of the first pieces of sequencing work in Argentina) Rodolfo Ugalde, Armando Parodi, Daniel Sánchez, Oscar Campatella, and Juan José Cazzulo.

Throughout this process, despite the growing number of groups, molecular biology research concentrated on a few areas. Other areas which were rapidly expanding on the international scene (cellular and molecular neurobiology, development genetics, structural biology, genomics and computational biology) had a low local development in Argentina. Among the most highly developed areas were the regulation of genetic expression, molecular parasitology (especially developments around *T. cruzi*), signal transduction, growth control and cellular differentiation, and traditional topics like protein glycosylation and the synthesis and decomposition of sugars and oligosaccharides.

During this period every group working on molecular biology was installed in laboratories belonging to universities (particularly in the Universities of Buenos Aires, San Martin, and Rosario) or in laboratories associated with universities (like the Campomar Foundation and INGEBI, both associated with the UBA). The only experience outside the universities was the creation of Institute of Molecular Biology (later transformed into the Institute of Biotechnology) at the National Institute for Agricultural Research (INTA). This institute was created by Eduardo Palma and, specially, by Ewald Favret, who was one of leaders of "traditional" genetics research during the 1960s and the 1970s.

Currently, studies on biology at the different schools of sciences are largely dominated by molecular biology, and it is difficult to identify a significant number of PhD dissertations not engaged in this field.

Concerning the institutionalization of the new field, it is important to mention that the Argentine Society of Biochemical Research, created in 1965 and headed by Leloir, changed its name in 1987 and became the Argentine Society of Biochemical Research and Molecular Biology.

Molecular biology also had a growing attention in terms of science policy instruments and funding. In a short period (less than fifteen years), public resources allocated to molecular biologists jumped from 0 to more than 20 percent of total expenditures financing scientific research (figures from National Agency for Scientific Research 1998 and 1999). Moreover, in 1982 the National Secretariat of Science and Technology (SECyT) created the First National Program for Biotechnology and Genetic Engineering, transformed in 1991 into the National Prioritized Program for Biotechnology. In addition, as we will see in the following section, almost all researchers in biotechnologies were (and at a large scale remain) molecular biologists or working on molecular biology topics.

The Shift to a Transversal Regime

Molecular biology's development as a discipline has brought intimate knowledge of genetic memory and its physical manipulation. The techniques used in this manipulation make up genetic engineering and allow us to extract genetic information from a human pituitary cell to manufacture growth hormone, transfer it to a bacterium (such as *E. coli*) and cause this to produce a biopharmaceutical: recombining human growth hormone. As Michel Morange (1994, p. 284) points out, "the development of molecular biology has, by integrating itself into other biological disciplines, given rise to a new biology, indeed, to a new reading of life itself". It is precisely this mode of intervention that has given rise to genetic engineering techniques.

From the mid-1980s, the development of molecular biology in Argentina shifted from the proliferation of research groups. On the one hand was a growth in applications of molecular biological techniques in other fields of knowledge, especially ones in which actors quite different from traditional, Campomar biochemists (agricultural research, pharmaceutical developments, genetic therapies) began to play a role. On the other hand there was the recognition that this new reading of life gradually acquired in such profoundly transformed traditional fields of knowledge as immunology, parasitology, and so on.

New research groups appear in this context, concentrating mainly on developing products by means of genetic manipulation techniques. The commercial use of some of these products is presented as a horizon of "application", and this gave rise to the Institute of Advanced Biotechnology at the National Institute for Agricultural Technology (INTA). The institute laid special emphasis on breeding transgenic plants such as Río Cuarto disease–resistant maize or new varieties of potatos. Elsewhere, there were developments in the breeding of transgenic animals (the genetic modification of mice at the INGEBI, again undertaken by researchers returning after their postdoctoral studies) and in the veterinary sector. Here, the Center for Animal Virology (CEVAN, dependent on the Science Council) is significant. Its molecular virology studies of foot-and-mouth disease and rotaviruses gave rise to a diagnostic "kit" based on monoclonal antibodies.

One of the most rapidly developing fields was the pharmaceutical industry and diagnostic reagents. Biotechnological laboratories sprang up, the most important of which was Biosidus. This lab has maintained close contact with research groups from the INGEBI and the Institute for Biology and Experimental Medicine (IBYME) in order to master certain specific techniques. The circulation of "academic" researchers to and from this laboratory has been widespread, making it virtually unique in Argentina.

Last, various research groups who keep their studies focused on the basics, and a growing mastery of third-generation biotechnology techniques have brought about profound changes in the way they investigate and define their interests. More widespread and better-understood techniques deriving from molecular biology and genetic engineering have consolidated the positions of research groups from the INGEBI and the Institute for Biotechnological Research (San Martín University), especially in the study of the *T. cruzi* genome.[42] More important still is the increasingly hazy dividing line between the various biological disciplines. This was particularly noticeable at the Campomar Foundation during the 1980s and 1990s, where differences between molecular biology pioneers and traditional biochemistry lines had fallen away.

Progress in this set of techniques has given rise to the development of advanced or third-generation biotechnology. Such techniques are generic research devices that give rise to a transversal regime of knowledge production. In other words, research agendas, as well as the social organization of work, are defined not by the particular framework of a discipline, but by its field of application and the type of product to be developed.

The development of biotechnologies was at the core during different public research policies, from the mid-1980s until today. Indeed, the set of institutional arrangements concerning biotechnology can be summarized as follows:

— 1985: Creation of the Argentine-Brazilian Center for Biotechnology (CABBIO).
— 1986: Creation of the Argentine Forum for Biotechnology.
— 1990: Creation of the National Commission for Agricultural Biotechnology (CONABIA).
— 1993: Creation of the National Commission for Biotechnology and Health (CONBYSA).
— 1999: Creation of the Permanent Commission for Bioethics and Biotechnology by the Chamber of Deputies.
— 1998–2002: Establishment of biotechnology as one of the high-priority areas by the National Secretariat for Science and Technology according to the biannual plan.

The Reconfiguration of Molecular Biology: Biotechnology and Genomics (1995–2015)

Research into *T. cruzi* is an extremely important aspect in the development of molecular biology in Argentina, a true coproduction of object and disciplinary field over several decades (Kreimer 2016b). Indeed, research on this subject focused on a significant shift in the 1990s manifested in the T. Cruzi Genome Sequencing Program, carried out by a number of different international laboratories. This crystallized in greater technification (the dissemination of automatic sequencers, scanning and digital reading, and so on), while certain continuities with more traditional research remained.

This new stage in the development of scientific knowledge certainly did not affect molecular biology alone, but rather most fields of knowledge. Its main features can be summarized as follows:

— Increased complexity of—and dependency on—equipment in almost all knowledge fields.
— Compared to the traditional laboratory, a parallel change of degree emerges in the unity and organization of the research problems, some of which are now more frequently tackled by "groups of groups" or networks involving a growing number of researchers.

— As a result, the costs of research tend to grow—and are often met by consortia, both public and in public–private partnerships— although there are also research processes and inputs that continue to be industrialized and that exponentially reduce certain traditionally high costs, such as genetic material sequencing.

— Changes in the most advanced countries' science and technology policies, organized in areas of influence promoting global networks with researchers from peripheral countries: the United States, the European Research Area (ERA), plus the strong emergence of China in recent years.

— Intensive use of information and communication technologies (ICTs) which enable remote virtual links and interactions in real time, with growing exchange and accumulation of shared information in databases (open or closed).

Against this background, we should mention the *T. cruzi* genome sequencing project, conducted between 1994 and 2005, and supported by the Special Program for Research and Training in Tropical Diseases (TDR) of the World Health Organization (WHO) and the Ibero-American Program for the Development of Science and Technology Program (CYTED), among others. This project was a true indicator of developments in the field of Latin American molecular biology, not so much because of its findings or contributions to intervention in the disease, but because it shows the new configurations of research.

Historically, the creation of the *T. cruzi* genome project follows the ability to physically map the human genome. Tools and approaches like bioinformatics and genomics, which only emerged in the late 1980s and early 1990s, played a key role in redefining the field, although they did not enjoy the same degree of prominence as they had in other parts of the world.

Though still the subject of debate and as occurred with the Human Genome Project (HGP), the results of the T. Cruzi Genome Sequencing Project (PGTc) fell far short of the pledges made, as sources of new approaches and expectations about the production of treatments for the disease, like vaccines, diagnostic techniques, or medications. Following the literature on the human genome, as well as interviews we conducted with actors linked to the PGTc, these exaggerated expectations stemmed from assuming an overall logic or emerging level in all genomic data (McKusick & Ruddle 1987) and from underestimating the time required by materialization in specific applicable results. However, both show an emergent form in the organization of scientific work, including a wider-ranging division of labor in terms of genome databases. In the case of the PGTc, the project's presentation highlights the project's technical characteristics rather than the issue's social relevance (see El-Sayed et al. 2005; Ferrari 1997).

The Role of Technical Change in the Reconfiguration of Molecular Biology: New Emerging Fields or Mutations Within the Field Itself?

To discuss the role of technical change in the reconfiguration of the local field of molecular biology, we need to review what it involved and why it more broadly altered the social and intellectual organization of biology between the 1970s and 1990s. During this period, molecular biology had already been given up for dead at least twice. The first time was in 1968, when Gunther Stent published the article, "That Was the Molecular Biology That Was"; the second, in the article "Towards a Paradigm Shift in Molecular Biology", by Walter Gilbert, published in 1991. Both men, who had been central actors in the field, had seen how molecular biology stabilized the new techniques and, therefore, lost its innovative character on entering ordinary technical life. As a result, this field felt in some kind of intellectual and disciplinary obsolescence. Morange (1994), for his part, described this rough ride during the 1970s as the "crossing of the desert". Beyond their dramatic tone, then, the diagnoses have in common a shared concern about the depletion of the intellectual and institutional foundations of molecular biology and the fact that those cycles of depletion and subsequent re-emergence were strongly tied to the technical changes introduced into biological research as of the mid-1970s.

By technical change, I refer—deliberately vaguely—to a succession of innovations in technical objects that began to form the experimental and theoretical life of biological research in the core countries as of the mid-1970s and which became intertwined with cognitive changes and social organization. We can reduce the innovations to two major groups: the first includes such innovations in biological material manipulation techniques as recombinant DNA-based methods, the use of molecular markers, or sequence copying through the polymerase chain reaction (PCR); the second, the complex processes of codification of biological material in the form of data and the tools for processing them. It includes, above all, bioinformatic software, genomic databases, and, more generally, digital technologies.

The important thing about the first set of techniques is that they enabled the intervention of the biological object at the level of individual elements of genetic material, supporting experimental practices and the continuation of molecular biological research without having being subsumed or traced back to the institutional and cognitive frameworks that still sought to monopolize biochemistry (Fox Keller 2003). However, whereas these techniques constituted the most qualified processes of scientific work within laboratory practice until the 1980s, by around the 1990s, manipulation techniques had begun to be standardized and were readily available through the supply of commercial kits. This process of industrialization of techniques represented a problem for many scientists. On a case-by-case basis, Gilbert feared that it would transform the molecular biology of the day into a technology more than a science: that is, a "pocket molecular biology" that could be practiced by "reading a recipe book or buying a kit" (Stevens 2013, p. 4).

The dissemination of these techniques was accompanied by a second process that reinforced the idea of depletion in molecular biology: *the inscription and subsequent mobilization of biological material in the form of data*. Putting it more synthetically, inscribing means the conversion of matter and energy into something written: in this case, the inscription included the manipulation of biological material at the level of individual units of genetic material and its subsequent migration in the form of numerical data. Sequencing represents a highly paradigmatic example of inscription: namely, determining the order of the individual components in the molecules of a cell's genetic material (the nucleotides making up the DNA) and subsequently representing them graphically or textually as a sequence of discrete data. From 1977 on, this process could be done in a relatively automated way and with the transcription of data to digital computer languages. Thence, the accumulation of research data began to increase exponentially, and the costs of sequencing dropped proportionally (Wetterstrand 2015).

Inscription processes are, naturally, no novelty in the inauguration of molecular biology. However, with the massification of molecular techniques in laboratories, the inscription process and subsequent data manipulation step came to be viewed as problematic for the practice and organization of biology. We can give two reasons for this here: tThehe first is the dissemination and industrialization of sequencing techniques, which, as with the previous set of manipulation techniques, sparked controversy due to the perceived opposition between molecular biology as a science and as a technique, respectively (Gilbert 1991; Lenoir 1999; Kaufmann 2004; Fox Keller 2003). The second reason is the "wave" of data generated with the dissemination of these techniques, with the possibility of inscribing these sequences relatively quickly, cheaply, and easily, molecular biology laboratories found themselves dealing with masses of data that were several orders of magnitude above what they thought they could manage.

In any case, the problem of the quantity of data was also a qualitative problem. How to manage the new wave of data? How to generate meaning from it in order to formulate and respond to the problems of biology? Who could deal with it, and what skills did they need? Would it still be the biologist? And if not, how they would relate to the biologists? The answer to these questions many times fell to computer science and knowledge engineering. However, the passage was not easy, much less direct.

The works of Hallam Stevens and Miguel García-Sancho (in the history of bioinformatics) or those of Stephen Hilgarter and Alain Kaufmann (on human genome sequencing initiatives) provide a detailed account of these difficulties: namely, the exchange between the worlds of biological cultures and informatics, conflicts, models of work coordination in laboratory networks, bids for publicity or privacy of results, speculation, and competition of scientists and their domain. Exchange and conflict between the domain of biology, on the one hand, and the representation and resolution of its problems in terms of data and information, on the other, is the aspect that seems to be most interesting when, later, we read the reconfiguration of the field in the context of Argentine molecular biology and the threats to its claims of autonomy during this last stage.

Informational discourse operated as a discursive guide to generate disciplinary proposals, directing ambitious research programs toward computerization and mobilizing vast resources in the service of global initiatives (Fox Keller 2002; Chow-White & García-Sancho 2011; Hilgartner 1995). One of the clearest examples of the crystallization of these expectations was the emergence of genomics and its subsequent connection with the design and development of the HGP.

A definition of genomics appeared in 1987 in the journal *Genomics*: Its proponents described it as "a marriage of molecular biology and cell biology with classical genetics . . . fostered by computational science". The purpose of this new discipline, they wrote, was to map all the expressed genes of an organism, regardless of their function, to obtain a "Rosetta Stone" to interpret the genetic mechanisms of diseases and their development. This same view accompanied the HGP a few years later, suggesting the advent of a shift from a "reactive" practice of medicine (one that cures the patient once sick) to a "preventive" practice (through the diagnosis of and subsequent intervention in defective genes) (McKusick & Ruddle 1987, pp. 1–2).

Although, since the last decade, the "emancipating" or determining role of genomics in the field of biology as a whole came to be answered: A degree of "disciplinary" differentiation currently prevails between the areas of genomics and bioinformatics, on the one hand, and a more traditional experimental biology, on the other, a differentiation expressed in professional profiles (training), practices (division of labor in research), and relatively well-defined institutional cultures and expressions. In the central countries, above all, this disciplinary differentiation and its expressions in distinctive working styles are more apparent and mutually reinforcing (Vos, Horstman, & Penders 2008; Hine 2006; Hilgartner 2004; Stevens 2013).

From this vantage point we can think about bioinformatics and genomics as a network of actors and practices with claims to autonomy or disciplinary demarcation of their own.

The Deployment of Change at the Local Level

We are interested in locating the noted changes in these new areas of knowledge and practices in Argentina in order to understand their role in this already highly structured (almost hegemonic) field of molecular biology. To do this, we pursue the perspective that views genomics and bioinformatics as results of coproduction processes: spaces where the world of information is bidirectionally linked to the world of life and gives rise to contingent results according to the particular contexts in which it develops. Jasanoff (2004) developed the idea of coproduction in order to analyze the interlacing between science, technology, and power. Here our ambition is more limited, and we pick up the proposal (already spelled out in works on computerization in biology) to distance ourselves from deterministic historical readings—social, political, or technological—or relativistic schemas within which a wider reading of historical and institutional processes is unimportant. Jasanoff suggests four essential pathways for further analysis in the "language" of coproduction; we rely heavily on the analysis of three of these: institutions, discourses, and representations. Excluding identities here, these pathways provide a view of relations between the traditions instituted in molecular biology and the changes and continuities of recent decades with the influx of genomics, bioinformatics, and sequencing projects, including their peculiarities, differences, and contingencies, as well as their connections with other contexts.

In spite of the cognitive and organizational ruptures that start to be introduced in the late 1970s and early 1980s, many aspects of the local biomedical tradition hinder the development of computational biology—anchored as it is in bioinformatics and/or genomics—as a disciplinary or institutional field of its own.

Furthermore, when bioinformatics tools and genomic approaches entered local biological and genetic research, they did so with a different dynamic to the one then being explored by central countries: It was the molecular biologists themselves who made contact with these new tools and methods and then introduced them into their own research, a practice that, by and large, still continues to this day.

Bassi, González, and Parisi (2007), for example, identified the development and growth of these fields with genome sequencing projects for different organisms. Still today, bioinformatics and genomics are particularly important in field research on Chagas disease, thanks to the results of the sequencing project of the entire *T. cruzi* genome. In fact, several Argentine researchers who took part in this venture played a fundamental role in the subsequent dissemination and implementation of genomic approaches and techniques in Argentina's science and technology institutions, while maintaining their more "traditional" role as molecular biologists.

The entry of bioinformatics and its relations with molecular biology in Argentina are also somewhat troubled. The first laboratory to bear the name "bioinformatics" was at the Institute for Research in Biotechnology of the National University of San Martín in the mid-2000s. Fernán

Agüero, who has been running the Laboratory of Genomics and Bioinformatics since 2006, had acquired his training in bioinformatics in 1999, when he went to Sweden to take a course before turning to research on *T. cruzi*. According to Agüero,

> working on my thesis, I began to train up in bioinformatics. That was in 1999. The main problem I had with my doctoral thesis was that the organism I work on is not an organism that can be grown in the laboratory . . . it has to be taken direct from infected animals. In this case, it was from cows, or sheep . . . not humans [. . .] the meatpacking plants were beginning to close, one plant after another, and we were running out of biological material. . . . Well, so I had no biological material to work on, so I went to Sweden to do the course on bioinformatics . . . until some material appeared [. . .] Then, tracking, I found that Ghiringhelli was teaching something, a course, a subject . . . at the time, I wasn't aware.
>
> (interview with Fernán Agüero, May 23, 2013)

It is not that bioinformatics tools and knowledge were nonexistent in Argentina: Since its inception in 1993, the PGTc itself had required the design and implementation of computing resources to store, analyze, and sort the sequences that make up the organism's genome. This development was led, in part, by Mariano Levin, of the Institute for Research on Genetic Engineering and Molecular Biology (INGEBI), in coordination with the Human Polymorphism Study Center (CEPH) in Paris (also linked to the sequencing of the human genome) and researchers from six other countries, as well as the CYTED network. Furthermore, as Agüero mentions, Daniel Ghiringhelli had been involved in the first formal approaches to genomic databases in Argentina. In 1993, he participated alongside Oscar Grau at the National University of La Plata in setting up a node—the first outside Europe—to provide access to European Molecular Biology Laboratory (EMBL) databases. Later, Ghiringhelli was linked to the first formal academic expression of bioinformatics by teaching the subject on the biotechnology degree course at the National University of Quilmes in 1999. Agüero, for his part, was the co-author of the first article on *T. cruzi*'s DNA sequencing, published in the journal *Genome Research* (Agüero, Verdún, Frasch, & Sánchez 2000).

The timing of these incorporations is extremely significant for at least three reasons: First, before the provision of commercial Internet connections (which only reached Argentina in May 1995), the use researchers made of computer networks was rather limited. Although the CONICET did have an Internet connection enabling institutional communications by e-mail as of 1986 (the Red CERCA), they had an extremely intermittent use until the second half of the 1990s. Moreover, these approaches to international databases (or even the generation of their own databases via the PGTc) were not ultimately viewed as "belonging" to genomics or bioinformatics; unlike what happened at the EMBL (and almost simultaneously in the United States), these areas in Argentina have experienced a slower, more difficult passage in stabilizing their practices and knowledge, and even slower and more difficult in their (still incomplete) autonomization as fields, compared to molecular biology and biotechnology. Finally, this period also marks biotechnology's transition to a transversal regime, in Shinn's terminology (2000), characterized by a central role for instruments and some melting of disciplinary identifications (Kreimer 2010d). The potential of biotechnology applications, for example, was one of the keys to incorporating genomics and bioinformatics as support tools for the research applied by Grau's team.

Toward 1990, by expanding the scope of the networks and the intensity of international collaborations, the division of scientific work within these networks was also expanded and reconfigured. In this sense, the 1990s continued the dynamic of proliferating molecular biology groups that had begun to take shape in the 1970s. Even so, the expansion of the groups does not form homogeneous spaces of knowledge production, but organizations that are highly segmented and not without tensions (Kreimer 2006, p. 204). Through this dynamic, two group profiles begin to form: on the one hand, those that are more deeply integrated in research projects and programs formulated in world centers of scientific production and having an active role (or frequent involvement) in international events (and capabilities to capture international resources); the second profile, on the other hand, corresponds to the rest of the groups, which enjoy little or no international integration, operate in relative isolation, and, understandably, have fewer opportunities to benefit from international funds.

The approaches, methods, and techniques from bioinformatics have played a variable role, depending on different points of view. In some cases, for example, bioinformatics was held up as a set of "auxiliary" techniques in the context of research done in more "traditional" molecular biology. In contrast, more recently, when local researchers directed their projects at research centers and international funding agencies, they needed to emphasize the role of genomics and bioinformatics in their research in order to ensure their projects met with a better reception.

Local bioinformatics in Argentina continues to have a far more dependent institutional and cognitive dynamic largely and subordinated to the field of molecular biology and biotechnology (Bassi, González, & Parisi 2007; Rabinowicz 2001). The merely incipient existence of institutional expressions and lower levels of disciplinary differentiation thus have their counterparts in the difficulty of formulating, through bioinformatics, research projects recognized by national evaluation and scientific funding bodies. And despite molecular biology retaining a degree of international hegemony throughout the field of the life sciences, genomics and bioinformatics in the central countries have managed to develop institutional and academic—and even cognitive—spaces more truly their own.

Whereas computational biology in the United States and Europe was a source of radical ruptures in how biology was done—part of an irreversible shift in the practices and organization of research—it seems to have come about far more quietly in Argentina. This means that change was—and still is—experienced as a pathway within biology, for which information tools operate as an essential technical means but do not in themselves always open up a field that deserves to be seen as their own in institutional or cognitive terms, as was the case with, for example, biotechnology (Kreimer 2010d; Rabinowicz 2001).

Since the technical objects in computational biology are growingly perceived as inseparable from ordinary scientific activity, the changes in this field are naturalized by the researchers belonging to other fields and research institutions and agencies. Looking toward the place of practices, this naturalization translates into an ambiguous transit through spaces and working styles, allowing individuals and teams to variably assume a more markedly computational, or else, more experimental profile. In the predominant approach in Argentina, researchers using computational tools are more often than not biologists with experimental training who oscillate between the two spaces.

One of the variables structuring the perception of their role is what generation researchers belong to. Depending on when they began to have contact with these tools and the timespans of their career development, researchers display four main ways of positioning themselves in their work:

1. Researchers who made contact with these tools at the late 1970s while doing their postgraduate studies or residencies abroad but who maintained a line of work based more on experimental practice of the "wet lab" type and on reductionist approaches. In this generation, the prevailing outlook is for computational approaches and tools to represent little more than a technical aid to biological research or, even more awkwardly, a science capable of formulating and answering their own research questions;

2. The second generation shares some characteristics with the first, such as postdoctoral training abroad. They are differentiated, however, by locally introducing the computer research tools they had made contact with abroad.

   Between the late 1980s and early 1990s, various initiatives appeared, usually connected to international initiatives to acquire resources or start-up projects linked to the genomic and bioinformatic developments being used abroad. Some of these initiatives involved relatively institutionalized projects explicitly oriented to genomics (the setting up of the EMBL node at the National University of La

Plata in 1993 by Grau and Ghiringhelli, or the first approaches to PGTc, coordinated in parallel by Levin and Frasch the same year), though there were also more isolated local initiatives, such as certain laboratories' efforts to computerize knowledge production. In this generation, the development potential of biotechnological and biomedical applications was closely associated with justification of the implementation of these techniques. However, the view of bioinformatics and genomics as a field in its own right was not manifested outside of the PGTc and had to wait for the next generation to settle into more stable configurations.

3. The third generation is still formed by biologists. However, these went one step further, incorporating computer tools and "silently" making the switch to computer-based approaches, but more explicitly developing the idea of formulating questions and foci on the basis of these new approaches.

   What we term "silent" here is, once again, the fact that they did not abandon an experimental biological profile while embarking on a more intensive use of computer-based approaches. In their 2007 description of Argentine computational biology, González, Bassi, and Parisi identified two types of groups in the country, depending on the academic background of their primary constituents: The first group was from chemistry and physics, with solid mathematical and statistical backgrounds, and has, in recent years, begun to turn its attention toward problems of biology; the other group had its origins in teams from biological disciplines that, in many cases, have been involved in genome sequencing projects and database curation (Bassi et al. 2007). The actors of this generation are, above all, biologists belonging to the latter type and have also created rather more institutionalized spaces for knowledge production through bioinformatics or genomics. Instead of replacing experimental biology, this generation of researchers leans toward the "dry lab"—computer science—with the intention of formulating new questions and problems through both genomics and molecular biology in the most "traditional" sense.

4. The fourth generation developed in the spaces opened up by the third. It consists of researchers trained in computational biology laboratories or doing their doctoral training in lines of work that focus on problems more clearly inscribed in bioinformatics.

   Although the spectrum of biological sciences continues to dominate their undergraduate training, they have the opportunity to take courses in bioinformatics during their undergraduate and postgraduate degree programs in Argentina. Qualifications aside, research profiles begin to appear that focus entirely (or mostly) on dry lab or in silico work; nevertheless, a significant portion of learning in these more clearly defined areas continues to be done informally and at their own expense.

Concluding Remarks

The entire period can be understood as a succession of transitions and slow generational renewal before, in a roundabout way, it reaches the foundation of a new, fairly autonomous field. The process of transition bears the following specific features.

First is the imitative nature of molecular biology arising from new lines of research, which we call subordinated integration. In other words, topical and conceptual innovation is due to the adoption of lines and techniques of research developed in laboratories belonging to the mainstream of international science through postdoctoral education abroad and a subsequent return to Argentina. Over time, this throws up networks of cooperation between "successful" local researchers and foreign centers of education. At the same time, it leads to a segmentation of local research groups into those that do manage to fit into these networks and others that do not.

Second comes the persistence of "crystallized" scientific traditions resistant to the emergence of new disciplines. With the "adoption" of lines of research already developed abroad, molecular biology as a "new discipline" plays an auxiliary technical role during a first transition stage. In a second stage, the discipline begins to consolidate itself as an autonomous field of knowledge with a growing number of research groups characterized by "topical ultra-specialization". This hampers participation in the conceptual issues behind the adopted topics.

This last feature has even given rise to rhetorical reconstructions by traditional researchers. According to this interpretation (which tends to play down the various schisms and watersheds), there ought to be a continuity between physiology, biochemistry, and molecular biology (genetics is not usually mentioned in this interpretation) in which each discipline represents a technical leap forward that does not derive from schisms in their ways of conceiving of the biological universe.

According to this form of conceptual innovation, the adoption of new problems, theoretical approaches, and research techniques that are not well developed locally encourages research practices with a heavy experimental bias (which confers reliability on any data obtained) and an intense concentration on a handful of research problems (topical ultra-specialization). This culminates in low levels of participation in general conceptual problems and leads to what we call a hypernormal science. Once the discipline has been "crystallized", it hinders conceptual innovation, and this is associated with difficulty in reinvesting the capital accumulated in new, uncertain frontiers of knowledge.

By the mid-1950s, the field of biochemistry was at its peak as a typical disciplinary space, internationally and locally. The essential principles of the paradigm had been around for several decades, and there was a whole "space of sociability" tied in with academic practices, local and international associations, specialized journals, and international events that constituted a powerful legitimation network. Biochemistry in Argentina had largely been established by Bernardo Houssay, the exclusive protagonist of scientific professionalization in the country, a central player in the development of physiology and self-appointed leader of a whole generation of researchers. Also involved were several of his students such as Luis Leloir (first at UBA and then at Bernardo Houssay's own Institute for Biology and Experimental Medicine and, after 1947, at the Campomar Foundation) and Alfredo Sordelli (at the Malbrán Institute), who had trained at the Buenos Aires Faculty of Medicine. Luis Leloir and Alfredo Sordelli aimed their research at the promising field of biochemistry. This field connected seamlessly with biomedical research and was seen by the actors themselves as continuous with the tradition founded by Houssay from the 1920s onwards. Molecular biology, however, had more complex origins. Here, various different disciplinary trends came together to generate a new field.

From the very outset, molecular biology exceeded the bounds of any one disciplinary field. It effectively organized itself as an expanded space comprising actors, traditions, approaches, practices, and forms of organization. Francis Crick (who, along with James Watson, had suggested the structure of the DNA double helix) is eloquent on this point:

> I myself was forced to call myself a molecular biologist because when inquiring clergymen asked me what I did, I got tired of explaining that I was a mixture of crystallographer, biophysicist, biochemist, and geneticist, an explanation which in any case they found too hard to grasp.
>
> (Crick 1965 in Stent 1968)

As I pointed out, the powerful hegemony exerted by the biochemical tradition in Argentina worked against the development of a transition regime. On the one hand, the rediscovery of genetics in the United States after Max Delbrück and Salvador Luria's work had no local correlate. By the 1950s, research into genetics had no deep roots in the academic world. On the other hand, neither was the "structuralist" school of thought, led by the John Bernal and his students, much developed. We have seen earlier how Cesar Milstein, who had trained in Cambridge with John Bernal's students, only spent a short time in Argentina at the Malbrán.

When the first molecular biology laboratories were dismantled in 1962, none of the figureheads of the Argentine biomedical tradition defended them. In fact, when Cesar Milstein turned to Luis Leloir for support at CONICET and, less explicitly, to sound out the possibility of settling in at Campomar, Luis Leloir took an indifferent attitude. In the end, Cesar Milstein decided to return to Cambridge.

This period in a transition regime is usually extremely trying for its practitioners: They do not obey the discipline's organization and legitimation criteria to the letter, while the representatives of the "mature" disciplines are not usually understanding (even viewing them with distrust or disinterest) towards researchers moving "on the periphery" of the established paradigms. The first laboratories in the bosom of the Campomar Foundation brought about a shift in levels of analysis and, especially, in the introduction of novel techniques. As such they had to obey two masters: On the one hand, as they were mostly young postdoctoral researchers, they had to make it appear that their works belonged to traditional biochemistry to keep the support of the local community figureheads. On the other hand, they aligned themselves with international trends that were consolidating a new field. There was no going back on the fact that, after the publication of *Phage and the Origins of Molecular Biology* in 1966 (Cairns, Stent, & Watson 1966), practitioners recognized a new field of knowledge.

By the early 1970s, research into the life sciences had also radically changed scale (some years after the shift shown by physics). As Gaudillière (2001) points out, the production of laboratory mice in the United States went from 65,000 in the 1930s to (according to Jackson Memorial Laboratory estimates) more than 35 million by the 1960s. For researchers who, like Luis Leloir, had been trained and had worked for years under the "little science" paradigm, such a change of scale was difficult to comprehend.

This transition regime began to establish itself by the late 1970s and was fully institutionalized by the time Hector Torres's group broke away from the Campomar Foundation in 1982 to found the Institute of Genetics and Molecular Biology (INGEBI). INGEBI was set up explicitly to take advantage of new research lines brought back by young researches on their return to Argentina. Scientists belonging to this new generation have experienced a substantive change: Most of them have studied biology, not medicine, chemistry, or biochemistry. Their training has socialized them with a set of techniques that have radically altered the old divisions of work, making it more horizontal and enhancing intralaboratory and interlaboratory collaboration. Levels of analysis have changed substantively too. By the mid-1980s, DNA sequencing had become almost routine, as had the belief that most significant biological processes operate at the intramolecular level. Lastly, as Michel Morange (1994, p. 284) has observed, molecular biology had gone from being a "science of observation" to being a "science of intervention", of action.

The next period broke with the transition regime in its extension of genetic engineering applications to various fields. Whereas the transition regime ultimately involved different spaces of disciplinary representation, the new forms of knowledge production brought in actors who until then had been "marginal" or "external" to the pure academic field. In the case of genetic engineering as applied to the transformation of vegetables, the incorporation of agricultural engineers, rural producers, seed producing companies, and public regulatory bodies made for an ever more complex space. Implementation, which had been important for molecular biology from the outset, now became central: Genetically modified seeds operated as true "genetic devices", used and resignified by different practitioners, in the way Gaudillière's production of mice were.

The development of biotechnologies is an excellent example of what Shinn has called "research-technology communities". In more developed countries, the emergence of these communities is quite clear-cut, but if we analyze the emergence and development of biotechnologies in Argentina, a peculiarity is immediately visible. First, the moment we move on to the level of technology, this presupposes a relationship between a producer and a user. Indeed, it is impossible to speak of a technology without a "user", though here the word does not necessarily mean the existence of capitalist enterprises based on current models. When we talk of "users", we refer to actors interested in the products in question, actors who establish some form of transactional relationship with the producer or producers.

So if, as I pointed out in previous sections, one of the characteristics of peripheral science is the inability to articulate the relationship between knowledge production and knowledge use (appropriation), how then do these relationships rearticulate themselves when this relationship stands at the heart of the development of a given technology, say, biotechnology?

The crucial thing in understanding this dimension is the emergence of new actors and/or the rearrangement of existing ones. According to Carlos Correa (1996, p. 27), in developed countries "we are witnessing the emergence not of a 'biotechnological industry'; but of a transectoral complex which uses biotechnology in the most diverse areas, concentrated in the industrialized countries and dominated by the big industrial and seed producing companies". This complex includes the existence of large-scale R&D laboratories devoted to experimenting with new procedures and products. It also includes constant interaction with more academic-minded research groups. There is a considerable flow of researchers from academic laboratories to company-based R&D groups. Poncet (2001) has put forward the interesting notion of "knowledge industrialization" in the life sciences in his analysis of the situation in France. He emphasizes the various transformations biotechnological knowledge undergoes before effectively being incorporated into a productive process. He points out that industrialization creates hybrid institutional spaces similar, to Terry Shinn's "research technologies".

Unlike what happens in industrialized countries, companies working in biotechnology in Argentina are generally small or medium-sized, and most of them are not using the most modern biotechnology. On the other hand, the number of academic research groups working on biotechnology-related topics has shot up in recent years. One only has to glance at the lists of subsidies granted by funding institutions over the last five years (in particular, the National Agency for the Promotion of Science and Technology) to see this trend clearly. The trend is even more visible if one looks at a longer period.

So, the birth of a technology apparently calls into question the construction of a user (or even a market). Is it therefore possible to view this growing mass of research from the academic sector as biotechnology? Only a partial answer can be given. First, one has to ask oneself about the (real or symbolic) "construction of the user" that gets researchers involved. Second, one has to look at how the relationship is established between researchers and other institutions (commercial or otherwise) bound with the appropriation of this knowledge. Third, it would be interesting to analyze changes in the organization of work, scientific configurations and, last but not least, the science–society relationship.

Certain empirical laboratory studies on the first of these problems have shown that researchers have directed their efforts at common topics and procedures in the central countries. Thus, one part of biotechnology research in Argentina appears to be close to the "frontier" of international knowledge. These topics (and products too) do not, however, seem to have incorporated the (real or symbolic) construction of a real, concrete user. An appropriate construction of this user is certainly not the only requisite for knowledge to be effectively appropriated by a significant social actor, in other words, for scientific knowledge to be incorporated into a technology. Effective appropriation relies on many other elements. Some of these depend on the strategies deployed by researchers, while others must be attributed both to context and to the strategies of other actors.

In fact, we observe different configurations over a half-century of developments in various scientific fields in Argentina, from the early and emerging experiences of molecular biology to the computerization of research and its possible reconfiguration as "information science". It is worth noting that, despite the actors' frequent discourse about "the objects reconfiguring and imposing forms of organization", the ways in which scientific practices and organization are articulated depend not so much on what might be imposed by the natural world, but on how actors define what is problematic or interesting in a succession of intersections that seem to be "natural", except during the periods of conflict or change.

Therefore, and with a certain resemblance to the Kuhnian approach, we are interested in placing the emphasis more on transitions and hybridizations than on periods consolidated, and thus naturalized, by the agents themselves.

As we have shown, the institutionalization of molecular biology as a new field, both internationally and in Argentina, faced severe obstacles.

At the international level, this was due to the convergence of divergent traditions coming from genetics, physics, crystallography, and biochemistry and meant that none of its practitioners wanted to abandon their previous spaces and venture into a new space with an uncertain future. Let us just say that the actors never know in advance the sociocognitive and institutional diversions that the emergence of new fields later makes apparent. Certainly driven on by the wide recognition obtained by questions like the discovery of the double helix or the role of the RNA messenger (both rewarded by Nobel Prizes), the actors of the day very gradually abandoned their fields of origin and espoused the new label of molecular biology. However, once this step was taken, their march was implacable: They "colonized" most of the research practices associated with the life sciences, reconfiguring almost every field, from mainly descriptive, taxonomic traditional biology, to research on plants and seeds, animal genetics, and, of course, biomedical research. In fact, once a gene can be autonomized from the organism it was taken from, its origin is of little importance, as demonstrated by the huge development of genetic engineering in the latter part of the twentieth century, pasting genes from one species into another and even synthesizing genetic material.

These traditions in Argentina did not have the force they had in England, France, or the United States. The main resistance came from the field of biochemistry, by then an established disciplinary space whose methods, despite making contributions highly valued by the international community (Leloir was awarded the Nobel Prize in 1970 for his work on the metabolism of sugars), were, in those days, rather precarious: Leloir himself boasted of manufacturing a refrigerated centrifuge from an old washing machine and an automobile tire filled with ice cubes. Against such a background, it was difficult to understand the changes taking place in research practices and, above all, to think that technical innovation is often the driver of large-scale conceptual innovations, or even disciplinary changes, as shown, among many other examples, by the building of huge particle accelerators, without which a whole range of issues could not even have been conceived (Galison & Hevly 1992).

It is striking that the phrase Leloir used to minimize the emergence of molecular biology as a new field was the same one used by certain molecular biologists forty years later to minimize the reconfiguration of their field by the introduction of a completely new computing paradigm: "it is nothing but a set of techniques". Apparently, once a new field is well established, it is more likely to accept adjustments to its founding theoretical frameworks (molecular biology had to deal with the serious questioning of its "dogma" and make the appropriate adjustments), perhaps more in the belief that conceptual adjustments end up strengthening the field's conceptual "hard core" than perceiving the potential risks of technological change in the reconfiguration of the objects, practices, social organization, and, ultimately, the disciplinary configuration itself.

The spread of ICTs in scientific research involves some important changes for all fields of knowledge and some specific restructuring in the case of molecular biology. In terms of practices as a whole, we note that a certain ideal formulated by Merton around the mid-twentieth century concerning the scientific community's scope for producing knowledge collectively, despite geographical or linguistic barriers, the mainstay of its ethos of collaboration and accumulation, might be easier to attain through the spread of ITCs: The circulation of knowledge has never been so fast as it is now; the ability to store information and make it available to one's peers has never had the power it has today.

What is more, the laboratory, the emblematic unit of knowledge production for several centuries now, is being increasingly challenged by collective ways of working that allow interaction—and even intervention—in real time from any corner of the planet. However, the availability of these techniques is a long way off from strengthening the democratic ideal, because access to knowledge does not respond to a pattern of uniform distribution, nor is the capacity to use it equally distributed. In a recent work, we show that over 40 percent of Latin American scientists participating in European networks are involved in information gathering and processing activities, whereas routine technical activities or equipment maintenance takes up 15 percent of their activities (Kreimer and Levin 2013.). In contrast, less than 10 percent of their activities is related to theoretical output, confirming our hypothesis about the consolidation of a new international division of scientific labor.

The dissemination of ICTs in molecular biology engendered a transformation and diversification of research groups. Some of them view genetic knowledge as having the same status as any other piece of digital information; in other words, there is no clear dividing line between bench work and computing operations. Other groups emphasize the division between "wet lab" work, the manipulation of living matter, and the subsequent digital encoding of information but keep the two processes as related but distinct spheres. As we are witnessing this development today, we are in no position to predict its outcome.

It is worth broaching one last question already mentioned in this chapter: namely, the uses of knowledge. Looked at in detail, every new transformation of the field (molecular biology) in Argentina has been accompanied with a set of promises about the generation of new, more complex knowledge that could, as well as overcoming the blinkered view of traditional perspectives, place more emphasis on its social use in tackling local problems. Despite the successive reconfigurations summarized earlier, the vast quantities produced and validated in the international arena, the knowledge actually used to address social problems has been minimal. Technological change and disciplinary reconfiguration are apparently not enough to change the peripheral cleavages in local science.

Table 5.1 Evolution of the Groups from the Area of Regulation Mechanisms

Group Heads	Members 1974	1975	1976	1977	1978	1979
Israel Algranati	M. Gracía Patrone G. Echandi Meza O. Burrote	M. Gracía Patrone G. Echandi Meza O. Burrone S. Goldemberg N. González	M. Gracía Patrone O. Burrone S. Goldemberg N. González	M. Gracía Patrone O. Burrone S. Goldemberg N. González L. Crenovich	M. Gracía Patrone O. Burrone S. Goldemberg N. González L. Crenovich	M. Gracía Patrone O. Burrone S. Goldemberg N. González M. Ferrer
Héctor Torres	M. Flawía H. Terenzi E. de Robertis (Jr.) N. Judewicz P. Leoni	M. Flawía H. Terenzi E. de Robertis (Jr.) N. Judewicz P. Leoni	M. Flawía H. Terenzi E. de Robertis (Jr.) N. Judewicz P. Leoni	M. Flawía N. Judewicz P. Leoni M.T. Téllez Iñon M.C. Maggese G. Glikin A. Kornblihtt	M. Flawía N. Judewicz P. Leoni M.T. Téllez Iñon M.C. Maggese G. Glikin A. Kornblihtt L. Molina y Vedia	M. Flawía N. Judewicz M.T. Téllez Iñon M.C. Maggese G. Glikin B. Kornblihtt L. Molina y Vedia J. Reig-Macia
José Mordoh (until 1977)	B. Fridlender G. Almallo	B. Fridlender M. Fejes E. Medrano	B. Fridlender M. Fejes E. Medrano	E. Medrano		
Romano Piras (until 1977)	M. Majeldeld A. Chepelinsky G. Daleo A. Horenstein	M. Majeldeld A. Chepelinsky G. Daleo A. Horenstein	M. Majeldeld A. Chepelinsky G. Daleo			

Source: Archives of the Campomar Foundation

Table 5.2 Researchers trained abroad

Researcher	Area	External Center
Héctor Torres	Regulation Mechanisms	Dept. of Molecular Biology, The Wellcome Research Laboratories, USA
Mirtha Flawía	Regulation Mechanisms	Dept. of Molecular Biology, The Wellcome Research Laboratories, USA
Norberto Judewicz	Regulation Mechanisms	IV International Course of Techniques of Molecular Biology School of Medicine, Washington University, St. Louis, USA
Eduardo de Robertis (Jr.)	Regulation Mechanisms	MRC Laboratory of Molecular Biology. England
Oscar Burrote	Regulation Mechanisms	V International Course of Techniques of Molecular Biology MRC Laboratory of Molecular Biology, England
Alberto Kornblihtt	Regulation Mechanisms	Sr. William Dunn School of Pathology, Oxford University, England
Manuel Garcia Patrone	Regulation Mechanisms	V International Course of Techniques of Molecular Biology
Armando Parodi	Glycoproteíns	Dept, Biologie Moleculaire, Institut Pasteur
Luis Quesada Allué	Carbohydrates	Embo Workshop on the Methodology of the Structure and Metabolism of Glucoconjugates

Table 5.3 Local Biotechnological Products on the Pharmaceutical Market in the 1990s

Product	Technology	Company
IFN-alfa	Cell Culture	BioSidus
IFN-r	Recombinant DNA	BioSidus
IFN-r	Recombinant DNA	Pablo Cassará
IFN-gamma	Recombinant DNA	BioSidus
Bovine SOD	Extractive	BioSidus
Insulin	Extractive	Beta
Human Insulin	Chemical	Beta
GC-SF	Recombinant DNA	BioSidus
HCH	Recombinant DNA	BioSidus
Timosine	Extractive	Serono

Acknowledgments

Many thanks to all the researchers I consulted, in particular Dr. Elie Wollman and Dr. Cesar Milstein, both key figures in this history, and to Professor Roy MacLeod for his helpful comments and suggestions. Part of this chapter was originally published as "Rowing Against the Tide: Emergence and Consolidation of Molecular Biology in Argentina 1960–90", authored by Pablo Kreimer and Manuel Lugones, pages 285–311, in *Science, Technology and Society*, Vol. 7 No. 2 Copyright 2002 © Society for the Promotion of Science and Technology Studies, New Delhi. All rights reserved. Reproduced with the permission of the copyright holders and the publishers, SAGE Publications India Pvt. Ltd, New Delhi.

Notes

1. There are many studies on the origins of molecular biology, including Abir-Am (2000), Gaudillière (1993), Creager (1993), and Mullins's classic (1972), among others. I have also found useful other texts, rather popularizing ones, like Thuillier (1975), Olby (1994), and Morange (1994).
2. As André Lwoff pointed out in 1966: "as the prophage is a molecule (of nucleic acid) and as I was studying its biology, I later became a molecular biologist. A fearful position, although around 1950 *would-be molecular biologists did not think of themselves like that*. The immeasurable virtue of the magic label was only discovered much later on. Perhaps I ought to add *that I am incapable of deciding to what extent I feel molecular, if indeed I am molecular at all*". See Lwoff (1966, pp. 88–99). The issue of equipment is particularly salient in that technical innovation has played a crucial role in the conceptual development of molecular biology. See Gaudillière (1996) and Kay (1993, 1996, pp. 446–447).
3. For a parallel analysis of this process in Spain, see Santesmases and Muñoz (1997).
4. UNESCO's policy was to set up National Science Boards in developing countries. This was particularly true of Latin America, where these types of organizations were set up (beginning with the CNPQ in Brazil) between the mid-1950s and the mid-1960s. However, as Feld (2015) pointed out, the adoption of a model for the national council was not a simple task. Indeed, several models and institutional features were in the core of the struggles from the 1950s on. For an in-depth analysis of this process, see Feld (2015). See also Amadeo (1978), Brawerman and Novick (1982), Marí (1982), and Oteiza (1992)
5. Houssay had founded in 1943 the Institute of Experimental Biology and Medicine (IBIME) with private, international backing, outside the university. In 1947, with the support of Jaime Campomar, Luis Leloir founded the Institute for Biochemical Research at the Campomar Foundation. Both institutes were key in the development of biomedicine and kept their private status, although later they were given dual "public-private" status and made dependent upon CONICET. See Cereijido (1990), Cueto (1994), Lorenzano (1994), Buch (2006), as well as the memoirs of the Campomar Foundation (1947–1984).
6. This issue has been discussed particularly in answer to "diffusionist" models of science, like Basalla (1967). See Lafuente and Sala Catalá (1992). Patrick Petitjean states that "in the establishment of scientific traditions [. . .]

'modernization' has not always been synonymous with 'Westernization', and the development of local dynamics, despite the influence of imperial domination, has depended largely on the ability of the local ´elites to draw this distinction". Petitjean (1996, pp. 7–11).

7. Houssay was the president, and most members of the board of directors were scientists with close links with him (many of them were his disciples), such as Leloir, Braun Menéndez, De Robertis, and Deulofeu. Pirosky was one of the few who, belonging to the biomedical field, had aligned with García (CONI-CET, Minutes of the Board of Directors Meetings, quoted in Feld (2015).

8. For details of this discussion, see Cereijido (1990). See also Varsavsky (1969), the introductory study to Varsavsky's works by Kreimer (2010b) and Feld and Kreimer (2012).

9. This led to an odd situation (similar to the French system), in which there were researchers paid by CONICET who worked in CONICET laboratories and scientists paid by CONICET who worked in university laboratories, as well as "pure" university researchers. CONICET researchers working at a university may or may not have had teaching responsibilities, but if they limited themselves to research, their salaries were paid entirely by CONICET. Those who did research and taught received part of their salaries from the university and part from CONICET.

10. The institute's origins date back to 1904. These involve the transformation of a group of laboratories in Argentina's Office of Public Sanitation into a Bacteriological Institute—dependent upon the National Department of Hygiene, which would later function as a center specializing in public health. For a description of the Bacteriological Institute's early years, see Estébanez (1996).

11. This journal was the *Revista del Instituto Bacteriológico Dr. Carlos Malbrán*, of the Department of National Hygiene. It was published annually from 1917. For the Malbrán Institute's activities in the early twentieth century, see Aquino (1921, pp. 35–36).

12. Cukierman (2007) suggests an alternative explanation. According to him, the creation of Oswaldo Cruz Institute responded to the concept of "disembarked science" which characterize the first period of Brazilian bacteriology.

13. An eminent "Pasteurian", André Lwoff (1981), points out that "Être pasteurien c'est donc appartenir à un ordre . . . Consciemment ou inconsciemment, les pasteuriens sont imprégnés par l'histoire qui les cimente, unis par la lutte qu'ils poursuivent pour la connaissance, par le combat qu'ils mènent contre la maladie, par la marche en commun vers un but intemporel qui s'éloigne lorsqu'on croit l'atteindre". (To be a Pasteurian is therefore to belong to an order [. . .] Consciously or unconsciously, the Pasteurians are impregnated by the history that binds them, united by the struggle they pursue for knowledge, by the fight they undertake against the disease, by walking together towards a timeless goal that moves away when one believes one reaches it.)

14. The procedure is called *Concurso* in Spanish. It is an open call to provide a given number of tenured positions in universities and other research institutions. In Argentina it is mandatory to organize this kind of open competitions to provide staff permanent positions.

15. In setting up the Virus Department in 1941, Sordelli managed to bring about an agreement between the Argentinean government (through the National Department of Hygiene) and the Rockefeller Foundation's International Health Division. By this agreement, the Argentinean government allotted special funds to set up the department, while the Rockefeller Foundation awarded scholarships to send two members of the institute to the United States to specialize in respiratory viruses. For further information, see Sordelli (1942).

16. Rudolf Kraus was the institute's first director, from 1914 to 1921. Kraus studied medicine in Austria and worked for several years in various German institutes. He was closely associated with the emergence of immunology and bacteriology in Latin America. From 1921 to 1928, Kraus was the director of the Butantan Institute in São Paulo, Brazil, and from 1929 to 1932, director of the Bacteriological Institute in Chile. For further information, see Aquino, op. cit. note 20. Continuing this tradition, Sordelli encouraged channels of communication among institutions in the biomedical sphere (such as the Argentine Biology Society, set up by Houssay), including exchanges of personnel (including Houssay's assistant, J. T. Lewis, who was head of the Malbrán Institute's pharmacology department between 1924 and 1928).

17. There is little data about the Institute under the Perón government (1946–1955). Our interviewees state unanimously that due to lack of funds, the institute went rapidly downhill. They also mention an indirect relationship between events at the institute and events at the University of Buenos Aires, when Houssay and Leloir and others had to resign their posts under political pressure.

18. Resolution No. 2.982 of 24 August 1956, which lays down the institute's organizational and administrative structure. This was reinforced by Decree Law No. 3283 of 26 March 1957, and its amendment, Decree Law No. 16145 of 9 December 1957. The following departments were also set up: Virology; Clinical Pathology (Serological and Microbiological Diagnosis); Biochemistry and Biophysics; Biological Product Manufacturing; General Bacteriology; General and Comparative Pathology; Protozoology and Applied Entomology; Pharmacy, Monitoring and Chemotherapy; Mycology; and the Genetics of Small Laboratory Animals.

19. This scheme was marked by a trade-off between the "Pasteur Model", where the head of each laboratory has a large degree of autonomy, and the "traditional" Malbrán Model, where the director always had great power over laboratory life.

20. The list was remarkable for those times: automatic autoclaves for sterilization, freezers, a lyophilizer for sera and vaccines, a BCG vaccine production unit, centrifuges, one large-volume centrifuge, Spino preparative ultracentrifuges, analytical centrifuge, a Servall centrifuge, a Tiselius electrophoresis unit, a variant paramagnetic resonance unit, a Beckman DK-2 spectrophotometer, a radioisotope unit, crosscurrent units, an electron microscope, a Defonbrune microscope, sets of Mettler scales, Leitz optical microscopes, automatic pipettes, potentiometers, homogenizers, chromatography apparatus, and a neurophysiology unit.

21. The list of scientists who emigrated is impressive: Cesar and Celia Milstein (Department of Biochemistry, Cambridge University), José Apelbaum (Instituto Di Patologia Speciale Medica e Metodologia Clinica, University of Siena), Julio Barrera Oro, (Baylor University, College of Medicine, Texas Medical Center), Manuel Brenman (Massachusetts Institute of Technology), Pablo Bozzini (Massachusetts Institute of Technology), Mariano Dunayevich (Department of Public Health, Viral and Rickettsial Disease Laboratory, University of California at Berkeley), Horacio Encabo (Faculty of Science, University of Paris), Ricardo Ferraresi (Pasteur Institute, Paris), Emanuel Levin (Department of Neurochemistry, Montreal Neurological Institute, and Department of Physiology of University College London), Antonio Lubin, (Communicable Diseases Center, Atlanta), and Jorge Raul Periés (Pasteur Institute, Paris).

22. The invention of the term "molecular biology" is attributed to Warren Weaver, program officer of the Rockefeller Foundation's Natural Science

Department. In 1938, Weaver went on record as saying that "among the research the Foundation is lending its support to, is a series belonging to a relatively new field, which might be termed molecular biology. This research uses subtle modern techniques to study the ever more minute details of certain vital processes" (Olby, 1994, pp. 616–619). The program supported fields of research rather than "individual scientific leaders", thus providing the means for general mobility. Abir-Am believes that "mobility factor" enables us to explain the appearance of molecular biology as a new sociocognitive network of connections within the main divisions of biological disciplines, physics, and chemistry, and therefore enables us to explain the multidisciplinary nature of molecular biology, unlike other disciplinary fields Abir-Am (1992, pp. 153–154).

23. The book *Phage and the Origins of Molecular Biology*, edited by John Cairns, Gunther Stent, and James Watson, was published on the occasion of Max Delbrück's sixtieth birthday in 1966. For an account of this story, see De Chadarevian (2002).

24. Pirosky entered the Malbrán Institute in the early 1930s as an honorary, unpaid assistant and carried out research on the fractioning of ordinary, antitoxic serums, specializing in the analysis of ultraviolet absorption spectra. Later, in 1935, he gained a position in the antitoxin and immunology department, and seven years later, he was heading it. He held this position until 1956.

25. The young chemistry major, César Milstein, speaks eloquently: "The Malbrán Institute was quite old [. . .] After a period of long neglect, it received a fresh lease of life with Dr. Ignacio Pirosky. The Government granted him special concessions, which allowed him to take on a great number of young scientists in full-time key positions. This was a bold and imaginative stroke but it placed Pirosky and the new young scientists in direct conflict with the sclerotic old guard. However, [during those years] an atmosphere of great scientific excitement developed among the recently hired scientists" Pirosky (1986, p. 29).

26. It is crucial to note that biology training at the university before 1960 was a very "traditional" (taxonomic) one, and most of the prestigious scientists working on biological research were medical doctors or biochemists.

27. The biologist, Pablo Bozzini, who went to MIT for postdoctoral studies and rejoined the institute in the early 1970s when its laboratories had been dismantled, was also a member of the department.

28. The first genetics institute belonged to the School of Agronomy and Veterinary Medicine of the University of Buenos Aires, headed by the agriculturist, Dr. Salomón Horovitz. The first degree courses in genetics were given by zoologist Miguel Fernández, who specialized in embryology, at the National University of La Plata, while at the University of Buenos Aires, the pioneer was Angel Gallardo, head of the Zoology Department in the School of Exact and Natural Sciences. See Vessuri (1994b, 2014) and Katz and Bercovich (1990, p. 82).

29. We refer to the work by Avery in 1944, which established DNA as the transforming principle of pneumococci.

30. The use of the phrase "internationalization strategy" derives from the hypothesis formulated by Abir-Am, (2000, 2002).

31. See Chapters 2 and 7. See also Vessuri (1996b), Gaillard (1996), Chatelin and Arvanitis (1990), Salomon (1994), and Petitjean, Jami, and Moulin (1992).

32. In addition to Wollman's course, a course on the problems of experimental design in biology and medicine and their interpretation was given by Dr. Ferting, the biostatistician, who came to Argentina in 1960. This responded

to Pirosky's explicit strategy: "From the moment I became Director of the National Institute of Microbiology, I believed that the way to ensure permanent scientific progress in microbiology and related sciences would be to maintain a constant, dynamic relationship with the most advanced centers in the world, not only through exchanging publications, but by inviting scientific figures to offer specialist short courses and sending grant holders to highly specialized centers" Pirosky (1986), my emphasis.

33. The equipment in the molecular biology department was valued at US$150,000, a large sum for the institute.

34. Milstein's words to the interim director on resigning speak volumes: "I am not aware of the selection mechanism that has been used [. . .] I have been passed over, though I am [. . .] the only person in a position to judge the relative importance of the work being carried out by the researchers in the molecular biology department [. . .] [Perhaps] you did not consult me because it is your opinion that all the work being done in this department is useless for the Institute. I must add that this interpretation is consistent with Minister Padilla's statement, in which case I personally feel I am one of those who are prompting the so-called 'misuse of state funds'" Milstein's letter of resignation, March 1962.

35. Pirosky was accused of handling resources in such a way as to have violated administrative rules. According to the (confidential) testimony of one researcher at the time, there were grounds for this accusation—not that Pirosky profited personally, but that there were "bureaucratic oversights" that freed up the administration of funds.

36. Due to the subsequent lack of specialist staff, the equipment in these laboratories was left in a storeroom until some of it was used more than fifteen years later. The rest was placed in storage (where it still lies today) and has become obsolete.

37. The group's line of research was based on the study of the mutations of the locus pyr-3nin the *Neurospora crassa* fungus. They were one of the first to use the fungus for genetic studies in Argentina. Reissig was one of the first to give the discipline any preliminary coverage in Argentina, with various advanced courses at the Faculty of Exact Sciences. Among those attending was Israel Algranati who, after working with Ochoa in New York, returned to the Campomar Foundation to found the first molecular biology group.

38. Institutional uncertainty again reared its head in 1976 under the military dictatorship, which forced the group into "internal exile". Only in the early 1980s did Oscar Grau (Favelukes's first student) consolidate the group's position institutionally to set up the Institute of Biochemistry and Molecular Biology at the National University of La Plata, which was closely linked to the National Institute of Agricultural Technology (INTA).

39. Ochoa had received the Nobel Prize in 1959 for his discovery of DNA polymerase, and as Santesmases and Muñoz (1997) point out, his "laboratory at the NYU Medical School was a scientific Mecca for the youngest Spanish biochemists and molecular biologists from the early 1960s". For a select group of Argentinian biochemists (and future molecular biologists), Ochoa played a similar role, we should add.

40. Interview with Dr. Alberto Kornblihtt, director of the molecular biology department at the School of Exact and Natural Sciences at the UBA (Buenos Aires: September 2001).

41. During the same period Oscar Grau formed the Biochemistry and Molecular Biology Institute of the National University of La Plata. Grau was a student of Gabriel Favelukes, one of the first molecular biology researchers in Argentina. However, the story of this group (not dealt with in this chapter) takes a profoundly different path from the Campomar groups.
42. For an in-depth description of Chagas disease and Latin American research on this topic, see Chapters 3 and 4.

6 Controversies on the Periphery

To Treat or Not to Treat (Chronic Chagas Disease Patients)?

Introduction

The development of Chagas disease has three clearly differentiated stages, characterized by specialists as "acute", "indeterminate", and "chronic".[1] In 1969, a scientific and medical controversy was sparked in Argentina and Brazil about the treatment to be given to chronic Chagas patients. Its resolution had—and is still having, given that the controversy is still raging today—direct consequences for the population affected.[2] Since then, a number of doctors have begun to use and test parasiticide drugs in chronic patients unanimously valued as effective in acute cases but of questionable and controversial effectiveness in chronic patients (Manzur & Barbieri 2002).

This chapter focuses on the development of this controversy in Argentina and aims to show the various stances of the actors involved (clinical doctors, cardiologists, scientific researchers, officials from the national state and international agencies), as well as analyzing the opposition of statements, arguments, and scientific knowledge. To this end, we identify four different moments in the controversy based upon (a) the amendments introduced by local and international agencies to regulations on the treatment of the disease and (b) changes in the stances and arguments of the local scientific-medical milieu regarding the use of parasiticide drugs in the chronic stage of Chagas disease. A first moment (1969–1983) runs from the emergence of the first cognitive disagreements to the passing of the first national regulations for treatment; the second (1983–1994) is characterized by the emergence of new scientific research and dissent and by the proclamation of the first World Health Organization (WHO) guidelines for treatment; the third (1994–2005) is shaped by the production of new scientific evidence and the second set of WHO regulations; and the fourth (2005–2007) is framed by the second set of national regulations and by the implementation and evolution of clinical research with international methodological standards.

We intend to show how the discourses and mechanisms that validate scientific proof are constructed. What forms the central object of dispute are thus the conceptions of disease and cure, both in the local medical scientific field, defined by academic publications and scientific congresses, and in the broader national and international health policy agencies.

It is important to note that patients have always been mentioned and involved in the controversy but have at no time actively participated in it. There are several reasons for this: For one thing, the fact that the disease particularly affects the rural poor, who are geographically highly dispersed in small towns, where the capacity for collective action is minimal; for another, along the same social and demographic lines, it is a population with scant access to either formal education or information about health issues.[3] Its institutional connections tend to be weak, and contact with national—and especially provincial—health officials is usually limited to such sporadic occasions as fumigations, for example. The third reason has to do with a "naturalizing" of patients towards Chagas disease (see Chapters 3 and 4).[4]

Chagas disease, then, is not thematized or perceived as a "problem"— or as something that demands intervention—by the patients themselves. Finally, the most important linkage for these sectors in terms of knowledge is physicians attending in hospitals, mainly cardiologists, many of whom are not even specialists in the pathologies associated with Chagas or even involved in the controversy (Sanmartino 2005).

As a corollary to this, there is a stark contrast with the participation of other social groups that have been constructed as a collective actor in relation to an issue perceived as a socially relevant problem; therefore, their modes of intervention—and the controversies that permeate them— are the subject of debate in the public arena. As an example, we can cite the collective mobilization of cancer patients when CONICET researchers claimed to have found new treatments based on South American rattlesnake venom (the drug crotoxin)[5] or the mass public and collective claims of AIDS patients for the new drug cocktails available to treat the condition in 1997 (Biagini, Escudero, Nan, & Sánchez 2005).

Controversies: Cognitive and Instrumental Dimensions at Stake

The study of scientific controversies has a long trajectory in the field of science studies, from the traditional historiographic approach to the studies enlisted in the constructivist perspective. Within the latter, Bath's school and, particularly, the studies of Harry Collins proposed to methodologically suspend the final result of a controversy (the "winner" position) and focus instead on how the social and cognitive consensus is constructed. This implies studying the debates on how the actors negotiate what is considered a *valid experience* and the diverse sense attributed by several practitioners taking part of a controversy.

The controversies have been, from the decade of the 1970s on, a privileged object in the attempt to develop a *symmetric* view of science; in other words, to show that, during the course of a dispute there is no distinction between "true" and "false" knowledge, but such category can only be established ex-post, once the dispute has been settled and a new consensus has been established. According to Callon and Latour (1990), the study of controversies was one of the most important ways no demonstrate the constructed character of scientific consensus, because they can show, throughout history, the so-called "science in the making" (*la science en train de se faire*). This perspective contrasts with the previous crystallized view of knowledge, according to which the result of a controversy was the triumph of true arguments ("winners") over false or inconsistent beliefs ("losers").

In methodological terms, the constructivist studies of controversies have been carried out observing the groups aligned with antagonistic positions, each one trying to impose its own viewpoint. These groups (positions) are defined at a specific moment when the controversy emerges, usually when a former consensus is broken or when a new topic arises. These positions, in general, are redefined throughout the time, until the closure of the controversy, when one of the groups succeeds in imposing a given perspective, or when, after a period of negotiations, a new consensus is achieved.

A controversy is defined by the debate between different positions about the validity of a body of scientific knowledge and the instruments serving to endorse it. Thus, the actors mobilize different arguments, discourses, instruments, institutions, etc., in order to strengthen their own position and to force the opponents to accept a given statement or fact. However, not all who take part in a debate have an equal degree of participation throughout the development of a controversy. Collins (1981a) says that those actors with a major role along the time form the "core-set"; they are also the actors for whom the result of the controversy is essential for their own practices.

A central concept, common to the different constructivist perspectives, is *negotiation*. Callon and Latour (1990), analyzing the emergence of constructivist analysis, suggest the resolution of a controversy must be thought as the result of several negotiations between actors. According to these authors (1990, p. 29), "the notion of negotiation brings a more accurate meaning than the notion of dispute", as far as negotiations refer to the *very content* of knowledge, which will, therefore, be modified during the negotiations through the commitments made by the different actors.

To grasp the development of the controversy, we need to remember that, between the late 1960s and mid-1970s, two transnational laboratories developed two parasiticide drugs—nifurtimox (Bayer) and benznidazole (Roche)—whose effectiveness was unanimously appreciated by the medical and scientific community as conclusive in acute Chagas patients, despite presenting several side effects. With regard to chronic patients, there were divergent, conflicting stances regarding the drugs' effectiveness because it was demonstrated that, with the provision of these drugs, patients were cured from a parasitological point of view (the disappearance of the parasite) but not from a serological one (antibodies were still present in the patients' blood) (Cerisola, Da Silva, Prata, Schenone, & Rohwedder 1977; Sosa Estani & Segura 1999). In 1983, these controversial results were behind the recommendation to exclusively treat patients with Chagas disease in the acute phase, both in Brazil and Argentina, while leaving chronic patients out of the specific parasiticide treatment (Sosa Estani & Segura 1999, pp. 166–170).

On the one hand, despite these treatment guidelines being issued, the controversy did not go away; on the contrary, the research aimed at testing the effectiveness of parasiticide drugs in the chronic phase of the disease grew. Moreover, this is also explained by the fact that, even today, the pharmaceutical industry shows no interest in conducting any kind of R&D activity to replace or improve either of the two existing drugs for the disease.[6] In fact, production of nifurtimox ceased around 1990 after a commercial decision by Bayer, which considered the medicine's consumer market no longer attractive. Roche made the same decision a few years later, announcing in 1999 that it was stopping production of benznidazole in Argentina. In the absence of a reaction from national laboratories or the Argentine authorities, the Federal Laboratory of Pernambuco (LAFEPE) [now the Pharmaceutical Laboratory of Pernambuco] in Brazil proposed to start producing it. After arduous negotiations, it was agreed that Roche would hand over the necessary expertise, though not the active principle, which is still produced by Roche at its plant in Basel, Switzerland, and subsequently exported to the LAFEPE.[7] Thus, the use of the existing parasiticides for groups of patients with no specific treatment—congenital and chronic Chagas—became, in the 1980s and 1990s, one of the main subjects of therapeutic debate and research among physicians working within the framework of the disease.

On the other hand, because the most important manifestations in the chronic stage appear as cardiac dysfunctions—"Chagas cardiopathy"—this research also coexisted with the widespread belief among many doctors that, whatever the origin of the cardiopathy, these were cardiac patients and should therefore be seen by cardiologists and given treatments available for that type of cardiopathy.

The Core of the Controversy

The core of the controversy is to determine whether or not it is appropriate to treat Chagas patients in the chronic stage of the disease with parasiticide drugs.

The first disagreements emerged with the results obtained by Brazilian researcher Romeo Cançado in 1969, who described the persistence of antibodies (positive serology) in chronic patients treated with parasiticide drugs, suggesting the continued presence of the disease (Manzur et al. 2002; Barclay et al. 1978). Later, between 1977 and 1978, the research results systematized by Cerisola and Argentine and Brazilian collaborators (Cerisola, Da Silva, Prata, Schenone, & Rohwedder 1977) regarding the effectiveness of nifurtimox in the chronic phase of the disease, and by Barclay et al. (1978) and Argentine collaborators about the effectiveness of benznidazole in the same phase, showed high cure rates among chronic patients from the parasitological point of view (the disappearance of the parasite from their blood). In this climate of uncertainty and vagueness, in 1983, the treatment guidelines drawn up by Argentina's Ministry of Health left out the parasiticide treatment for chronic patients and prioritized serological criteria over the parasite in determining a cure. Despite, or rather because of this, the medical controversy in Argentina did not end, and research continued to be developed aimed at testing the effectiveness of the drugs on the chronically ill.

Throughout the development of this controversy, the knowledge that defines the diagnosis, evolution, and cure of the disease were constantly called into question, as were the instruments to measure and describe the three states of Chagas. The legitimacy of the various practices of evaluation and validation of treatments were also questioned: In other words, the various methodological parameters that coexist within the framework of clinical research on the disease.

With respect to the diagnoses and the determination of a cure, there are currently three different, sufficiently stabilized criteria:

1. Parasitological: defined by the absence of parasites in the blood.
2. Serological: defined by the absence of antibodies in the organism.[8]
3. Clinical: a decrease in cardiopathy and an improvement in the patients' quality of life is observed.

Each criterion has a corresponding diagnostic technique: xenodiagnosis, serology, and, in the latter case, mainly an electrocardiogram and general evaluation made by the attending physician, respectively.

Each technique emerged and became widespread as a routine method at different points of the twentieth century, so it is worth briefly summarizing this historical development to show which methods and beliefs about the diagnosis, evolution, and treatment of Chagas became established among doctors, researchers, and officials.

As we have seen in Chapters 3 and 4, Chagas disease has been a frequent object of medical and scientific research from different disciplines for almost a century, since it was discovered by Carlos Chagas in Brazil. At the same time, throughout the twentieth century, it has become a public topic: Its extension, seriousness, and social and economic range impelled the active participation of the state, and several types of programs have been held in order to fight against the disease (Zabala 2010; Kreimer & Zabala 2007).

The scientific research about Chagas disease began in Argentina in the second decade of the twentieth century, when Salvador Mazza headed the creation of the Society for Northern Regional Pathology, in the Province of Jujuy, later transformed into the Mission for the Study of Argentinean Regional Pathology (MEPRA), dependent on the University of Buenos Aires.

In the early days, under the guidance of Salvador Mazza, between 1926 and 1946, in the MEPRA 1,244 cases of Chagas were identified using microscopy, animal inoculation, biopsy, xenodiagnosis,[9] and the Guerreiro-Machado reaction. Microscopy was a parasitological technique that directly observed and analyzed the parasite, while xenodiagnosis and animal inoculation detected it belatedly and indirectly (Mazza 1949). This set of techniques was particularly effective in the detection of patients in the initial acute phase, given that, at that stage, there are abundant parasites in the blood. The Guerreiro-Machado reaction is a serological technique—one that detects antibodies—and its usefulness lies in identifying patients with the chronic infection, when the parasites are not easily detected in the blood.[10]

The main biological and clinical features of the acute stage of the disease were then described and characterized (Segura 2002). Some years later, Cecilio Romaña carried out epidemiological research in order to establish the extent of infection areas at the Institute for Regional Medicine Regional (IMR), created in 1942 at the National University of Tucumán. Romaña aimed at establishing the relation between the infection with the parasite and the development of cardiac affection, identified as cases of chronic forms of the disease, and tested new different diagnosis techniques. Later, from 1947, he investigated the prophylaxis actions against the disease (Zabala 2010). During those years, work was done through the IMR on improving the Guerreiro-Machado reaction, also called complement fixation reaction (CFR), utilized to identify chronic patients.

In 1952, during Perón's government, a new institutional frame was created to perform both research and patient care activities: the National Service for the Prophylaxis and Fight against Chagas Disease (SNPLECh), ruled by the Direction of Transmissible Diseases of the Ministry of Health. The creation of the SNPLECh implied the establishment of a new public state institute devoted to knowledge production and patients' attention exclusively devoted to Chagas disease. Both tasks had been carried out until then by university institutes.

In 1957, after the fall of Perón's administration, the facilities of the SNPLECH were restricted to a laboratory installed in a plot of land ceded by the Institute of Sanitary Entomology. This laboratory was under the direction of the medical doctor (surgeon) José Alberto Cerisola and started producing *Trypanosoma cruzi* crops and *vinchucas*, with the main objective of improving the diagnostic test (Segura 2002, p. 51). On the other hand, a doctor, Mauricio Rosenbaum, trained as a cardiologist, was appointed the new director of the cardiology service of the SNPLECH in 1957 (Elizari 2003). Between 1956 and 1957, Cerisola and Rosenbaum carried out a series of epidemiologic studies that adopted the name of "Survey about Chagas disease" (Rosenbaum & Cerisola 1957); their goals were to systematize the cardiologic clinical symptoms of Chagas' chronic stage and to estimate its prevalence in different regions of the country. The information produced by these surveys had a very important role in characterizing and legitimating the conception of the Chagas chronic stage as a cardiac affection.

In the 1950s, at the National Service for Prophylaxis and Resistance against Chagas Disease (SNPLECH), epidemiological surveys were conducted that showed the relationship between the development of specific heart disorders and the presence of the *T. cruzi* parasite as a typical sign of the chronic disease. They also tended to institutionalize a new technique for routine diagnosis—the electrocardiogram—and to calculate the number of people infected on regional and national scales, which had previously been done haphazardly without discriminating different areas of incidence. The widespread routine use and application of the electrocardiogram in epidemiological surveys was central to the production of such knowledge, which would later stabilize as a new diagnostic technique alongside traditional xenodiagnosis and serology, which continued to be used complementarily.

At those times, the methods used to estimate (that is to say, extrapolate) the number of infected patients from specific samples were not developed enough, and the number of affected people was clearly overestimated (Zabala 2010). Probably this was influenced by the fact that the actors involved in the problem had to publicly show a large number of cases—even it would seem somewhat exaggerated—in order to draw the attention of the authorities to the "public problem" (Kreimer & Zabala 2006). In this way, a "percentage" of the infected patients that could develop Chagas disease was estimated, but this percentage was not really known. The same low precision affected data on the clinic manifestations of the disease, whether they be nervous, cardiac, or gastrointestinal. At this point, the "polymorphic" character of the disease, established by Carlos Chagas in 1909 and somehow since 1957 present in the 1940s, was displaced by a hegemonic conception of the disease as a basically cardiac ailment, conceptualized as *chronic Chagas myocardiopathy* (Rosenbaum & Cerisola 1957).[11]

Obviously, performing epidemiologic surveys had consequences in terms of health policies: Three years later, many mechanisms arose to deal with the disease not only in the national but also in the international context. Thus, several actions were planned in relation to detecting, diagnosing, and treating affected people: the obligation to notify the cases of infection, to take samples from the soldiers, and the creation of the National Program Against Chagas Disease. In 1963, the first regulation establishing the compulsory test to detect the presence of Chagas infection in the blood of donors was established. From this year on, following an order by the army, the serological test with the antigen of *T. cruzi* was added to the medical tests of young men joining the compulsory military service (Zabala 2010).

In the same year, the Panamerican Health Organisation (PAHO) selected the Sanitary Laboratory, headed by Cerisola, as the Argentinean representative in a multicentric research organized to standardize Chagas diagnosis techniques (Segura 2002). In 1966, the Sanitary Laboratory Dr. Mario Fatala Chabén[12] was transformed into the Institute for Diagnosis and Research of Chagas Disease "Dr. Mario Fatala Chabén" (INDIECH 1995).

The INDIECH implied an important innovation with respect to the traditional (and only) way that existed until then to fight against and to prevent the disease: spraying rural houses with insecticide. In fact, between the late 1960s and the mid-1970s, clinical tests on humans were conducted at INDIECH to evaluate the effectiveness of two drugs against *T. cruzi*: Bay 2502 or nifurtimox (Lampit Bayer) and, a few years later, benznidazole (Radanil Roche). As a result, administrating these two drugs was approved, but only for acute patients. The chronic patients were left out of the pharmacological treatment and have been treated, since then, just as cardiac patients. Thus, the symptomatic manifestations have been treated independently of the causes that originated them.

However, these techniques were not enough to account for all cases—or all stages—of the disease. In recent years, as a result of the process of "molecularization" in the biomedical sciences as a whole, a fourth method has been developed, based on detection of the presence—or absence—of DNA fragments of the *T. cruzi* parasite in blood and tissue. It is performed through the widespread polymerase chain reaction (PCR), which is more sensitive than traditional diagnostic techniques.[13]

The legitimacy of the various methodological parameters coexisting in the framework of clinical research on the disease is not a given, but brings us back to the way clinical research on parasiticide treatment in chronic patients is institutionalized. For some decades now, so-called randomized clinical trials (RCTs)[14] have been widely used in the most advanced countries, typical of evidence-based medicine (EBM).[15] They are considered to be the most legitimate international standard of maximum legitimacy, allowing the formulation of health policy recommendations with greatest regulatory force. However, reproducing such studies under these standards in peripheral contexts is no easy task: It entails having a community of doctors and patients involved in systematic, daily research. This is a high-cost (among other things, because of patients' mobility, availability, and the repeated interruption of their performance at work) and firm commitment to the research on the part of the patients and an infrastructure where it is possible to perform the monitoring and systematic attention of a significant number of patients.

These conditions were hard to maintain for groups doing clinical research into Chagas disease in Argentina. In fact, there are only two systematic works on the effectiveness of the treatment with trypanocidal drugs in chronic patients conducted within those parameters: the Treatment in Adults, or TRAENA, and the Benznidazole Evaluation for Interrupting American Trypanosomiasis, or BENEFIT, both done at the Fatala Chabén National Institute of Parasitology (INP). With the exception of these two research ventures, the works that prevail in Argentina are (a) "retrospective" studies, which use the clinical data available (on patients already treated), estimate developments, and make comparisons with untreated groups; and (b) "transverse" studies, which analyze the prevalence of certain pathologies at a given time, without retrospective analysis or prospective follow-up.

It is worth inquiring into the reasons why the development of clinical research with standard protocols was problematic within the framework of research into this disease, because, in principle, the existence of a strong local biomedical research tradition, alongside the robust development of the various branches of medicine, seemed to provide a suitable framework for the adoption of these guidelines. However, a plausible hypothesis explaining the absence of these studies in the framework of Chagas research should leave the weight of that tradition aside and put first the industrial pharmaceutical laboratories' historical lack of interest in this disease (with the exception of the first—and only—experiments conducted by Bayer and Roche several decades ago) due to the fact that it mainly affects poor, rural social sectors and internal migrants.

As we can see, the controversies' foci and points of interest vary with the different actors that take part. So, if we believe that the effectiveness of the studies being performed can only affect the resolution of the controversy insofar as they are accepted by the most significant actors, the "real" effect of the drugs on chronic patients is just one of the topics up for dispute. Another issue is how to establish the validity, or institutionalization, of pharmacological clinical trials on standardized bases, which then act as a true "test" in any future disputes. In this way, it is not just the "knowledge" that it is sought to legitimize, but the very practices of clinical researchers, whose institutionalization, unlike that of "basic" research (these concepts must be read as "native" terms; that is to say, those used by the subjects), still lags some way behind in Argentina. We examine the details of this issue in the next section.

The Actors and Politico-Institutional Devices Linked to the Controversy

Since the controversy developed, and primarily since the first national treatment guideline in 1983, local physicians researching and treating

Chagas began to adopt the following stances toward parasiticide treatment in chronic patients: (a) pro, (b) con, and (c) no active stance (but not providing the treatment in the chronic phase).

The "pro" group is noted for conducting clinical research designed to test the effectiveness of the treatment in question. This group comprises Rodolfo Viotti's team, based in the Chagas Disease Section of the Cardiology Service of the Eva Perón Interzonal Hospital of Acute Patients in Buenos Aires Province; Elsa Segura and Sergio Sosa Estani's group in Dr. Mario Fatala Chabén in Buenos Aires; and Diana Fabbro's group at the Research Center for National Endemic Diseases (CIEN) of the Faculty of Biochemistry and Biological Sciences belonging to the National University of the Littoral (UNL), in Santa Fe Province.

The three groups have very different institutional roots: the first, a hospital environment; the second, the area of public health research; and the third, the university. Each of these areas has its own quite distinct institutional mission and social function: While the university is defined as an area of original knowledge production and teaching, the environment of public health research produces knowledge by drawing up research agendas based on national health problems set as priorities from the point of view of the health policy of the population, and, finally, the hospital environment is oriented to providing professional prevention, care, and support services in order to solve health problems occurring a day-to-day basis.

This divergence of institutional goals is linked, in turn, with each area having been shaped historically by specific, divergent research traditions and working cultures which affect different cognitive practices: research activities per se and service/assistance activities (in the case of hospitals and public research, the proportions tend to the latter type of activity, whereas in universities the opposite tends to be true). Associated with the differential weight of these practices in each of these areas, it is possible to identify substantive differences between the type of knowledge produced, the forms of allocation of prestige and construction of authority, the circuits of knowledge accreditation and validation, the type of user/public to which it is oriented, the kind of construction of knowledge utility, the socioprofessional careers, and other elements. As an example, in the case of universities, the product into which knowledge usually crystallizes—and whereby it becomes legitimate—is scientific publication, which also assumes that other scientists are the primary users. In the other two cases, while the paper does have a presence, the paramount thing is products like diagnostic kits, vaccines, and so on, oriented to users such as the Ministry of Health, other medical professionals, or the patients themselves.

The main arguments mobilized by this group have been, on the one hand, that, while the decline or stabilization of cardiac symptoms does not imply absolute cure, it does produce an improvement in the patients' quality of life; on the other hand, while retrospective studies have no legitimacy when it comes to producing recommendations, they are valuable from the clinical point of view (Schapachnik 2002; Viotti et al. 2006).

Notable in "Group 2"—the "cons"—for his active participation in the controversy are Rubén Storino, a member of the Argentine Cardiology Society, and Sergio Auger, a doctor from the Cardiology Service of the Santojani Hospital in Buenos Aires. The main argument mobilized by this group has been to criticize the fact that "the cure is defined solely according to parasitological and serological criteria, while the clinical criterion is rejected", because the research geared to observing "the efficacy of parasiticide treatment in chronic patients is not performed according to international standards" (EBM). Other reasons cited by them have been that, in the chronic stage, the parasite is found in tissue (no longer in the blood), where parasiticide drugs are ineffective. Therefore, it seems logical to attack the heart lesions, not the parasite. Thus, they conclude, "it is not possible to determine the effectiveness of drugs due to the uncertainty over the cure criteria".[16] They add to this what they see as a sensitive issue: that the drug has too many adverse side effects to be used in patients where the risk–benefit ratio is uncertain (Storino 2002). Finally, there is "Group 3", which, without taking an active, public stance one way or another, does not usually treat chronic patients with parasiticide drugs. This group contains clinicians and medical cardiologists who treat patients exclusively in hospital services and conduct no research.

The positions for and against the treatment of chronic Chagas patients can be schematized as follows (Tables 6.1 and 6.2).

These different stances produced differing definitions of cure and disease that were neither only restricted to nor developed in the field of medicine or biomedical research, but spread to other institutional arenas where the cure criteria for the different stages of Chagas disease had to

Table 6.1 Positioning on Treating Chagas Chronic Patients ("Pro" Group)

Those who claim to be *in favor* of the treatment recognize that the retrospective studies are not valuable for recommendation making, but they consider them to have legitimacy due to the improvement of a considerable part of the patients from the clinical point of view.
Those who claim to be "against" the treatment, consider that the research orientated to observe the effectiveness of the parasiticide treatment in chronic patients is not carried out "with the method of medicine grounded in evidence", thus it has less validity as a "test method" and so it lacks "scientific character", which is conclusive at defining their position.
So let's see schematically the arguments for and against treatment effectiveness. Among those who are in favor stand out those who believe that the treatment with parasiticide drugs:
1. Minimizes, prevents, or relents the evolution of the cardiopathy.
2. It is capable of curing the recent chronic disease in children and adolescents as in the acute cases, and of curing, in a small percentage, the chronic disease (Schapachnik 2002; Viotti et al. 2006).

Table 6.2 Positioning on Treating Chagas Chronic Patients ("Con" Group)

Those who are against the treatment assert that:
1. The research made with the intention to prove the effectiveness of the treatment is not carried out with MBE method and therefore it lacks scientific basis.
2. In the chronic stage of the disease, the parasite is found in the tissues (not in the blood) and the parasiticide drugs have no action at this level.
3. In the chronic stage, what must be treated are the cardiac injuries—and not the presence of the parasite—which may be approached, as any cardiopathy, from the regular practices of cardiology.
4. It is not possible to determine the effectiveness of the drugs due to the uncertainty of criterion of cure
5. The drug at stake has many adverse effects to be used on patients when its effectiveness is uncertain (relation risk–benefit) (Storino 2002).

be negotiated and settled. At the national level, there was involvement from the Ministry of Health and the National Administration of Drugs, Foods and Medical Technology (ANMAT) and, at international level, the WHO (through several technical consultations conducted in the region). In these spaces, they confront the doctors who have been developing research lines for more than ten years, intended to test the effectiveness of parasiticide treatment in chronic patients, with doctors based in the Argentine Cardiology Society (SAC) and its Chagas Council, who are against such treatment. In collaboration with WHO, National Ministry of Health, and ANMAT technical staff, the two groups of physicians produced different documents which, over the last twenty years, have discussed, established, and amended the criteria to define and apply the diagnosis, care, and treatment for Chagas disease. These documents were crystallized in three national guidelines[17] and, at the international level, two technical reports of the WHO.[18]

The concepts of "cure" and "disease" have evolved over the years, depending on the arguments deployed by different actors. Taking into account the changes in the positionings, interventions, and types of argument mobilized, we can identify four different moments in the controversy: a first moment (1969–1983) bounded by the emergence of the first cognitive discrepancies, up to the sanction of the first national regulations for treatment; a second (1983–1994) by the emergence of new scientific research and dissent, and by the proclamation of the first WHO guidelines of treatment; a third (1994–2005) by the production of new scientific evidence and the WHO second guidelines; and a fourth (2005–2007) by the second national regulations, and the implementation and evolution of clinical research with international methodological standards.

These moments were identified using two criteria: one, the amendments that international and local agencies introduced in the regulations to treat the disease; two, the local medical environment's shifts in stance and argument regarding the use of parasiticide drugs in the chronic stage of Chagas disease.

First Moment (1969–1983): The First Disagreements and the First National Regulations for Treatment

In 1965, clinical evaluation trials began on nifurtimox for the treatment of acute and chronic Chagas infection in Brazil, Chile, and Argentina. The first results obtained by researchers were analyzed at a meeting in Santiago de Chile in 1968, where it was reported that "the treatment of acute Chagas infection produced a high cure rate with serological negativity (absence of antibodies)" (Cerisola 1977). As mentioned earlier, the first disagreements emerged with the results obtained by the Brazilian Romeo Cançado and other researchers in 1969, who described the persistence of antibodies (positive serology) in chronic patients treated with parasiticide drugs, suggesting the continued status of the disease. For their part, Cerisola, and his Argentine and Brazilian collaborators presented the results of their research conducted between 1977 and 1978 into the effectiveness of nifurtimox in the chronic phase of the disease. At the same time, Barclay and others, working on the effectiveness of benznidazole in this phase, demonstrated high cure rates in chronic patients from the parasitological point of view (the disappearance of the parasite from the blood).

In 1983, the guidelines for treatments developed by the Ministry of Health excluded parasiticide treatment for chronic patients, prioritizing the serological criterion to establish the cure, and recommended the implementation of drugs only for acute and congenital cases of Chagas.

Second Moment (1983–1994): The Emergence of New Research, New Dissent, and the First WHO Regulations

From 1983, new scientific research work emerged in Argentina which debated the issue and endorsed the parasiticide treatment for chronic-stage Chagas patients.

Some researchers suggested redefining Chagas as an autoimmune disease, observing that even in the absence of the parasite, the signs were still present in the tissues (Sosa Estani & Segura 1999).[19] Instead, other researchers argued that, while there were autoimmune components, the disease was ultimately caused by the action of the parasite (Tarleton & Zhang 1999). This issue continued to be a part of the debate, since no unanimous consensus has been achieved.[20]

We can sum up the two sides of this debate as follows:

1. The parasites cause an imbalance in the regulation of the immune system, promoting the expansion of polyclonal B cells, with consequent

hypergammaglobulinemia and the appearance of autoantibodies (Hontebeyrie-Joskowicz & Minoprio 1991).

2. The autoimmune origin of chronic Chagas is due to molecular mimicry, supported by evidence of the existence of antigenic determinants shared by parasite and host (Levin et al. 1990).

In this context, there was little scope to discuss the provision of a parasiticide drug, whereas the preeminence of the autoimmunity theory aimed to move the discussion onto another level: The disease is caused by the presence of the parasite, but its development is due to the fact that the immune system becomes the primary aggressor for the organism (interview with Rodolfo Viotti, Higa Eva Perón Hospital, June 21, 2005). From this perspective, it made no sense to administer drugs whose effect was precisely to kill the parasite.

At the international level, the WHO conducted a Technical Consultation in 1991, which recommended not treating patients with benznidazole during the chronic phase given that "no information was available about the effectiveness of the treatment in preventing the development of the disease" (WHO 1991).

The Argentine Cardiology Society (SAC)—particularly the members of its Chagas Council—opposed treatment with benznidazole in the indeterminate and chronic phases, actively intervening in the controversy without completely sharing the autoimmune hypothesis or expressly dwelling on it. The society's position has been an extremely important source of scientific legitimacy and authority for professionals who treat chronic patients as "just another kind of cardiac patient".

Third Moment (1994–2005): New "Proofs" and the New WHO Regulations

Although the debate on the autoimmune issue continues to this today (Bonney & Engman 2015), its intensity had waned by the mid-1990s. In 1994, the first results were published intended to show the efficacy of benznidazole in the late chronic phase (Viotti et al. 1994). In this work, the authors described "a significant decrease in antibodies and the development of minor electrocardiographic disturbances in patients treated with benznidazole as compared to untreated patients" (PAHO/WHO 1998). Other researchers (Ferreira 1990; Galvão, Nunes, Cançado, Brener, & Krettli 1993) had shown that

> if follow-up on patients stretched over a long period (more than five years), some of the patients treated in the chronic phase who had antibodies during the first few years after treatment started, had negative serology (the antibodies disappeared) in much longer periods (10 to 20 years).

> (PAHO/WHO 1998, p. 5)

These works suggested that a section of Chagas patients responded to treatment regardless of the stage of the disease or their age. The variability of the results depended on patients' monitoring times, which was measured in months for acute patients, years for early chronic patients, and decades for late chronic patients (PAHO/WHO 1998).

The Higa Eva Perón Hospital group's publication revived the controversy over parasiticide treatment in chronic patients. This was when, more explicitly, in publications, forums, and congresses about the disease, stances were more actively expressed "against" the effectiveness of this treatment due to the scientific invalidity of such research, given that, among other things, it lacked EBM standards (Viotti, Vigliano, Armenti, & Segura 1994).

Four years later, in 1998, a new technical consultation was conducted by the WHO. It concluded that there was no age limit to indicate treatment and left the decision to the discretion of the attending physician.[21] This proposal was based on "available evidence" about the relationship between the parasite and myocardial inflammation, on the regression of cardiac lesions with the specific treatment, and on the demonstration that treatment could reduce the occurrence or progression of cardiac lesions evaluated by electrocardiogram (Elizari 1999).

In the same year, at the national level, the treatment guidelines were amended, and it was recommended to include among the patients to be treated adults in the latent phase or with some "incipient or asymptomatic cardiac pathology". However, it was recommended to exclude Chagas patients with chronic organic manifestations from treatment (Ministry of Health 1998).

Fourth Moment (2005–2007): New Guidelines, Toward a New Consensus?

In 2005, the Argentine authorities produced new guidelines for the attention of patients infected with *Trypanosoma cruzi*, based on a revised draft of February 1998. Its most substantive modification lies in re-establishing the fifteen-year age limit for treatment to be administered. Beyond this age, it was left to the discretion of the attending physician (Ministry of Health 2005).

According to the testimony of researchers from the Fatala Chabén National Institute of Parasitology (INP), who take part in the TREANA and BENEFIT projects I mentioned earlier and who collaborated in the process of developing these new guidelines, the fifteen-year treatment limit is attributable to new strategies being deployed by research groups working to "produce new scientific evidence" in accordance with state-of-the-art methodologies on medical therapeutics (EBM) (interview with Dr. Sosa Estany, August 2006). This strategy thus reflects confidence in the future results of the ongoing TREANA and BENEFIT studies, conducted under the standards of the EBM paradigm.

Therefore, if all the actors were to accept the results of these studies as forming "proof", or otherwise, of the treatment, the controversy would be well on the way to being settled and a new consensus constructed. In other words, studies can only operate as "proof" in the resolution of the controversy if the group of cardiologists historically opposed to the treatment of chronic Chagas patients were to accept these parasiticide results and subsequently to consider them as a legitimate fact, as something firmly established and shared. The central object of the controversy, therefore, seems slightly displaced: It is no longer only or mainly about observing the effects of drugs on chronic patients, but about calibrating the value of clinical trials without much of a tradition in pharmacological research into the disease in Argentina.

Final Considerations: The Consequences of the Debate

The analysis of the present controversy charts the movement and interference of scientists not only in the space of scientific research into the disease but also in the field of its management as a public health problem defined by the Argentine Ministry of Health, and in the space of international recommendations on the treatment of the disease outlined by the WHO. There is a status of the researchers that is both political—when they act as experts and consultants—and scientific: In the WHO's case, its Technical Reports are the result of a large amount of research. In the ministry's case, they combine research activities with consultancy tasks and intervene in the design of the Guidelines for the Treatment of Chagas disease.

In particular, we have shown that, in the third moment of the controversy, the drafting of the National Treatment Guidelines and the WHO Technical Report was modified to include aspects linking to and deriving from the research results of local medical groups. Although such modifications did not recommend the use of treatment in chronic patients, they did relax the time periods for indeterminate cases. The WHO "left to each doctor's discretion the issue" of how to set the age limit for treatment. This may be read as an expression of the inability to form a stable consensus accepted by all. More room was also made for the clinical perspective (a criterion based on the clinical evaluation of patients) in the understanding of the disease that had prevailed until then. For its part, the Ministry of Health recommended treating adults in the latent phase or with emerging pathologies but excluded chronic patients with organic manifestations.

Both recommendations meant very different things to the actors involved in the controversy and acted as different signals for the legitimacy needed by medical professionals to provide specific treatment in chronic patients. The fact that most medical professionals did not provide treatment and would not revise their stance on this issue because of the 1998 WHO amendment was due to their more direct commitment and legal responsibility under national public health regulations.

It is worth emphasizing a particular feature of "peripheral" contexts mentioned in previous chapters: In the field of health—though it may apply to other fields as well—WHO recommendations are usually taken as a "black box" and applied in Latin American countries almost without debate. However, several authors have pointed out very emphatically that "we found that many strong recommendations issued by WHO are based on evidence for which there is only low or very low confidence in the estimates of effect" (Alexander et al. 2016).

Indeed, besides other restrictions, a structural weakness of most Latin American countries is its low quality of public decision-making, rarely supported by robust locally produced knowledge. Jasanoff (1998) has developed the concept of "regulatory science" as a distinct domain of scientific production, accountable to epistemic as well as normative demands, in ways that help explain its vulnerability to challenges from both science and politics. Indeed, regulatory science is extremely weak in Latin America and very often knowledge produced in developed countries or international organizations—like the emblematic case of WHO—tends to be taken as the basis for public decision-making (Kreimer 1996).

It is interesting to differentiate three types of professional profile (the third of which contains two subtypes), all actively involved throughout the controversy (following Collins' terminology, they form the "core set" of this controversy). The first is cardiologists who treat patients but do no systematic research; that is to say, although they can collect information about the patients and the disease, they do not meet the MBE parameters Their primary workplace is the doctor's office and hospital services. In this group, we have to differentiate between "generalist" cardiologists (those treating Chagas pathologies, among many others) and those who have concentrated more on the treatment of Chagas pathologies. A second profile consists of clinical researchers, who divide their activities between patient care and systematic research with well-defined parameters of legitimation. These researchers share their work, say, "between stretchers and laboratories" often located in their own hospital units. The last professional profile is identified with professionals engaged in research into the disease. Here, we find two subtypes differentiated by their institutional belonging and the type of knowledge they are expected to produce, which functions as a device to differentiate different circuits of legitimation: first, those located in universities or CONICET research institutes performing academic research, whose knowledge product is usually scientific publication; second, those belonging to public research institutes, like the Fatala Chabén INP, combining research activities with care tasks and health services. The knowledge produced is consequently objectified and coded not primarily in paper format but in a product such as a vaccine, serum, diagnostic kit, or epidemiological map.

We can assume that these different professional profiles are not outside of the public stance on "what is at stake" in the development of the controversy: For those who have to orient their practices about how to treat sick people, the decision whether to prescribe antiparasitic medication or not substantively modifies their own practices and even doctor–patient relationships. In contrast, for those researching the action of the parasite on the body, it is not just a question of orienting relations with patients—if they perform care tasks—but of the very representation of the knowledge on certain biological entities, the functioning of the immune system, and other similar topics.

It is, moreover, interesting to note that doctors often officiate as "mediators" between patients—and their blood or plasma—and scientific researchers, though, of course, they do not play a neutral role in this process.[22] As in any negotiation, there is a degree of give-and-take: If, to anchor their position, doctors require studies that use PCR (a technique not commonly available), they must seek such tests from lab researchers, yet instead they become, for example, suppliers of the material needed by scientists.

To summarize, the analysis of this controversy serves to illustrate the interrelatedness and differentiatedness of the scientific space against the broader field of society and how scientific consensuses are maintained and negotiated by actors that are not exclusively—or even primarily—scientific. We have also seen how the role of technics (scientific instruments or machines) has a central place not just in the production of the observable in scientific research but also in the definition of what is meant by cure and (parasitic, autoimmune) disease (Helden & Hankins 1994). These criteria have gradually changed, depending on the reference diagnostic technique used (xenodiagnosis, serology, PCR). This has consequences that go beyond the strictly scientific space, giving rise to the legitimation of certain medical practices and infrastructure in the attention and treatment of acute and indeterminate patients, but not for chronic ones.

While the controversy is not yet closed, most of the actors from the medical and scientific community have taken the line that the disease should not be treated with parasiticide drugs. The practical consequences of this majority stance at the level of practical attention and treatment have, then, resulted in most chronic Chagas patients not being medicated with specific drugs, but treated as a specific case of "cardiac patient".

Notes

1. Some researchers believe there are two, not three, stages: "acute" and "chronic". According to them, the chronic stage is subdivided into two further stages, "indeterminate" and "chronic" (Barclay et al., 1978). See also Manzur and Barbieri (2002).
2. Although the controversy originated in Brazil, where research on the subject was conducted with the participation of medical actors and Brazilian experts in its field of definition, and the WHO, and remembering that the controversy's evolution in Brazil is not the object of this work, what we can say is

that we do not see the activation and dynamism that the controversy had in Argentina in terms of local publications, public stances by national medical federations and associations, or consequences for the treatment effectively given to chronic patients.

3. In contrast, in the case of AIDS, since its origin, communitarian groups, patients associations, NGOs, and activists participated in the scientific, technological, and international medical debate (Epstein, 1996; Rabeharisoa & Callon, 2008).

4. An example of this kind of "naturalizating" the disease was mentioned in Chapter 4, where officials from the (national or provincial) Ministry of Health visit populations at risk and ask them what diseases are prevalent there. The most frequent responses are often a short list of diseases, like measles, chickenpox, flu, and so on. When the official asks them "No Chagas, right?" the answer many times is "Ahh . . . yes, Chagas . . . Everyone here has Chagas" (Sanmartino & Crocco, 2000).

5. For a summary of the "crotoxin case", see Braun (1989). For an analysis of the social mobilizations and belief structures around the case, see de Ipola (1997).

6. There are numerous reasons for local and international pharmaceutical laboratories' lack of interest in developing a more effective drug for Chagas disease. But it is clear that the nature of the market—a rural population living in conditions of poverty—and the relative absence of governments go a long way to account for it. For a more detailed analysis, see Kreimer and Corvalán (2010)

7. According to authorities at the Roche laboratory in Argentina, they proposed to local authorities to carry out this transfer with the participation of national (public and private) laboratories but, receiving no reply, decided to open negotiations with the LAFEPE. However, the authorities of the Fatala Chabén Institute, specializing in parasitology and a benchmark of Chagas research, claimed they only learned of the situation—the business decision and transfer—after the agreement with the Brazilian institute had been closed (Kreimer & Corvalán, 2010).

8. Antibodies are glycoproteins (proteins attached to sugars), also called immunoglobulins. They are secreted by a particular type of cell—plasmocytes—and are highly reactive with molecules called antigens. Antigens can induce the formation of antibodies (each antigen is defined by its antibody, which interact by spatial complementarity). Plasma cells are the result of proliferation and differentiation of B-lymphocytes. Their function is to recognize foreign bodies like bacteria and viruses and keep the body free of them. The production of antibodies is part of the humoral immune response.

9. This method consists of reproducing the natural cycle of the parasite in laboratory conditions in provenly negative triatominae which are fed the patient's blood. Boxes of nymphs are used, obtained from laboratory hatcheries. They are fed on bird's blood and made to fast for two weeks prior to application. The areas recommended for placement on the patient are the forearm and calf. The insects' feeding time during the procedure should be approximately thirty minutes. The removed boxes are kept in the insects' ideal breeding conditions, namely, a dark place at temperatures of 25°C to 30°C (Storino, 2000, 2002; Storino & Milei, 1994, p. 345).

10. In the presence of *T. cruzi* antigens, a specific antibody is able to form an antigen–antibody bond capable of fixing complement through Fc fraction of the immunoglobulin. This invisible bond is revealed by an indicator system that allows the free complement—that is, lacking antibodies—to produce

hemolysis in the hemolytic system, of ram's red corpuscles and hemolysin. Preparation of this technique is complex, its success depending on the parasite's antigens losing their duly standardized lipids and on the constituents, complement, hemolysin, red blood cells, white blood cells, and so on, having been duly tested in relative amounts (Storino & Milei, 1994, p. 349).

11. The authors use the concepts "miocardiopathy" and "Chagas chronic cardiopathy" indistinctively.

12. In 1963 the name of "Fatala Chaben" was added to the old Sanitary Laboratory to honor a researcher of this laboratory who died due to a myocarditis provoked by *T. cruzi*. He accidentally infected while he was carrying out his experimentation.

13. PCR was not routinely practiced in diagnosing Chagas disease, as it was suited to research techniques in molecular biology and was carried out in scientific research laboratories rather than hospitals and health centers.

14. Therapeutic clinical research based on RCTs is organized according to a number of formalized elements in a research protocol: a methodology with clearly defined research stages; methodical selection of the study population, and the formation of the sample type and number according to explicit, previously defined inclusion and exclusion criteria; external supervision of the results by a committee of experts not participating in the protocol; randomization of the sample organized on the basis of a test group receiving the therapy under study, and a control group given placebos or some other therapy; and a double-blind reading, where neither patients nor researchers know if they are in the test group or the control group (García & Teira Serrano, 2006).

15. EBM draws together a variety of meanings: "The conscientious, explicit, and judicious use of current best evidence in making decisions about the care of individual patients [. . .] including an orientation toward critical self-evaluation, the production of evidence through research and scientific review, and/or the ability to scrutinize presented evidence for its validity and clinical applicability [. . .] it mainly denotes the use of *clinical practice guidelines* to disseminate proven diagnostic and therapeutic knowledge" (Timmermans & Berg, 2003).

16. The uncertainty over cure criteria alludes to the coexistence of different cure diagnosis techniques mentioned in the previous section. The stances against bring the use of PCR into play in the controversy, claiming that, in many cases under study, its use yielded positive results in patients with negative xenodiagnoses. This showed that, despite the absence of parasites in the blood, the disease existed at tissue level. According to these results, classical approaches to cure—xenodiagnosis and serology—overestimated the medicine's real efficiency. According to these stances, herein lies the uncertainty of the cure criterion, because PCR is not, for now, used routinely in studies on the effectiveness of parasiticides in chronic patients

17. These are the Health Program Secretariat Resolution, Ministry of Health and Social Action and COFESA (1983); Manual for the Care of Patients Infected with Chagas Disease, Ministry of Health and Social Action (1998); Manual for the Care of Patients Infected with *Trypanosoma Cruzi*. Ministry of Health and Social Action (2005).

18. These reports are Chagas Disease Control, Report by a WHO Experts Committee (1991); Etiological Treatment of Chagas Disease, Conclusions of a PAHO/WHO Technical Consultation (1998).

19. In someone with an autoimmune disease, the immune system mistakenly attacks the cells, tissues, and organs of its own organism, and the immune system becomes an aggressor toward the very organism it should be protecting.

20. More recently, Bonney and Engman stated that "Although questions regarding the functional role of autoimmunity in the pathogenesis of Chagas disease remain unanswered, the development of autoimmune responses during infection clearly occurs in some individuals, and the implications that this autoimmunity may be pathogenic are significant" (2015, p. 1,537).
21. On the basis of these documents, amendments were made to the recommendations provided in the previous consultation, in 1991. As goals for the treatment of the chronic stage, it was also proposed "to eradicate the parasite, prevent the onset or progression of visceral injuries, and interrupt the chain of transmission" (PAHO/WHO, 1998).
22. For a discussion about intermediaries and mediators, see Latour (2005). He talks about the relationship between *mediators* and *intermediaries* as both human and nonhuman objects: *Mediators*, unlike *intermediaries*, "transform, translate, distort, and modify the meaning or the elements they are supposed to carry" (p. 39).

7 Centers and Peripheries Revisited—Internationalization of Latin American Science

From "Bricoleur" Scientists to the International Division of Scientific Work

Introduction

I intend here to present and explain an emerging model that I have called a "new international division of scientific work" and its consequences for more advanced Latin American countries. In the first part, I provide a historical outline of the internationalization of sciences in Latin America in order to show that, since the last quarter of the twentieth century, we are facing a new configuration. This new scene implies that, if until the last two decades the relationships among "central" research groups and those located in "peripheral" contexts left to the latter a little *marge de manoeuvre*, these relations have taken from then on the form of a closed or "take-it-or-leave-it" approach, marked by the emergence of "mega-networks". Thus, elite researchers from nonhegemonic countries are increasingly invited to take part in international research consortia, but the conditions to access are even more strict and the room for negotiations tends to be minimal. To illustrate this configuration I will present two particular cases: environmental research in the Patagonian area and plasma physics on the "periphery of the periphery". These subjects have been chosen—among many others—because they are considered "hot" in the current international science and thereby give several keys to understanding the new trends.

* * *

Internationalization has been present from the very beginnings of research in Latin American countries. Indeed, the institutionalization and development of "modern" scientific fields—particularly in the late nineteenth and early twentieth centuries—were closely tied with local researchers' relations with the disciplines' leaders in Europe, on visits to Latin America by these "travelers", or stays abroad by Latin Americans. This marks a first stage, which I term "founding internationalization".

Once disciplines were established in local institutions, the nature of researcher relations shifted: The setting of research agendas and conceptual innovations take place against a background of local–international tension. This corresponds to a second stage, which we can call "liberal internationalization".[1] Following the end of World War II, while science and technology (S&T) policies were being established in most developed countries—and cooperation protocols in science and technology were being institutionalized as a result—international relations became more formal and more institutionalized, marking a stage of oriented "liberal"[2] internationalization.

However, over the last quarter of the twentieth century the nature of these relations shifted yet again. If, in earlier stages, negotiations between researchers from the centers and peripheries left the latter a little room for maneuver, a new trend emerged, with collaboration taking the form of a closed, take-it-or-leave-it approach. This stage is marked by the emergence of "mega-networks" (incorporating 500 or more researchers) and "research regions". A new dynamic now became visible between hegemonic and peripheral groups. There is a paradox here: Elite researchers from nonhegemonic countries are increasingly invited to take part in international consortia, but the conditions of access are ever stricter and the room for negotiation tends to be minimal. It is worth noting that the new international division of scientific work (Kreimer 2010a; Kreimer & Meyer 2008) I'll show in this chapter often takes the form of subcontracting agreements (of course, with the mutual consent of the contractors), which in one sense is analogous to—and at the same time different from—the delocalization of certain industrial activities.[3]

1. From Liberal Internationalization to the International Division of Scientific Work

A Scheme of the Stages of the Internationalization of Argentine Science

To gain a better grasp of the challenges now present in an analysis of the dynamics of current knowledge production, I believe it is essential to direct our attention towards the historical role of international dimensions in the institutionalization and development of various scientific fields. The aim is not merely to place the research process historically as we might analyze it today: It is, above all, a matter of drawing attention to the fact that in most cases international relations played a crucial role in the origins and constraints of local scientific traditions. Far from involving a purely institutional or formal dimension, such relations had important consequences for the nature and content of local research agendas, for the topics developed and, most notably, for the scientific styles set in motion within each discipline.

From the institutionalization of the first disciplinary fields (by around the last quarter of the nineteenth century) we can identify four different stages in the socio-institutional and cognitive development of scientific research, adopting the structure of international relations for each period as a criterion for classification. This can be summarized as follows in Table 7.1.

Table 7.1 Stages in the Internationalization of Argentine Science

Stage	Main features	Period (approx.)
"Founding internationalization": from colonial to "national" sciences.	Institutionalization of new scientific fields. Visits by European experts (later also from some U.S. experts, depending on the discipline).	1870–1920
"Liberal internationalization" (Phase One): collaborations with the center.	Local leaders and "do-it-yourself" scientists: individual negotiations of research agendas with central teams on mainstream topics.	1920–1960
"Liberal internationalization" (Phase Two).	Towards "big science": emergence of S&T policies in Latin America and development of research support instruments. Shift to foreign postdocs.	1960–1990
New "international division of scientific work"	Integration into broad networks and mega-science: negotiation by Argentine researchers almost zero.	1990–

Throughout our period it is the cognitive control of scientific activities that is at stake, although it takes on very different forms in each stage, as we will see shortly. To put it briefly, if in the first stage European "visitors" establish the topics that organize the institutionalization of a new disciplinary field, which are presented as universal (and therefore conceal their local origin), in the next stage "local leaders" have to negotiate their topics and methods with the leaders of the hegemonic groups aiming to be recognized as fully fledged researchers within the "core set" of each disciplinary field. In the last stage, once the disciplines are well established in most Latin American countries, the hegemonic agendas and topics are controlled through relations between the leaders of the hegemonic groups, international or supranational agencies, and private companies located in developed countries. The leaders of peripheral groups are only invited into this process later on to carry out sometimes highly sophisticated tasks, the technical, cognitive and, above all, conceptual definition of which is quite out of their hands.

National Sciences

The origins of the first disciplinary fields were, as I have pointed out, closely linked to journeys by Europeans and Americans, particularly in the fields of physics and astronomy, respectively, as a number of examples clearly demonstrate.[4]

The first Argentine astronomical observatory was founded in Cordoba in 1871, under the direction of Benjamin Gould, the U.S. astronomer and creator of the *Astronomical Journal*. He had trained at Harvard University and—significantly—had followed his studies through with Carl Friedrich Gauss in Göttingen, Germany. It was at his request that the state purchased the first instruments, previously nonexistent in Argentina: chronographs, photometers, and telescopes (Rieznik 2008). His successor at the observatory in 1885 was another American, his assistant, John Thome. Unlike Gould, Thome was closer to French lines of work, linked to physical astronometry, which was viewed as archaic compared to German astrophysics. The observatory's third director, appointed in 1909, was Charles Perrine, another American, who ran the place until 1936 and tried to modernize the old lines of research and to institutionalize astrophysics (Bernaola 2001). It was not until Perrine retired that the first local director, Juan José Nissen, was appointed. In his speech on taking up the post he declared that "under Thome's direction the main building has been used for some time as the headquarters of the United States Consulate in Cordoba" [and that] "until 1936, no Argentine astronomer had worked there" (*ibid.*). Nissen, who held a doctorate in Mathematics from the University of La Plata, had studied mathematics, astronomy, and physics in Italy and Germany, which clearly marks an epochal shift, involving the first "national" leaders completing part of their training abroad.

The case of the institutionalization of physics is in some ways analogous. The University of La Plata's Physics Institute was set up in 1906 under the direction of the German physicist, Emil Bose. The university's authorities sought a candidate from among the German physicists of the day and decided to travel to Germany to persuade Bose, who had been director of the Danzig Institute of Technology and a student of Walter Nernst, to come and organize the new institute in Argentina (Bibiloni 2005). Bose took personal charge of equipping the first laboratories—the first of their kind in the country. This involved purchasing apparatus for experiments with alternating and continuous current and assembling the battery, accumulators and switchboards, the air compressor, and the liquefiers. As in the case of astronomy all the equipment was acquired abroad. In this instance it was Bose himself who traveled to Germany to buy it; he also took care of taking on the German technicians and experts for the installation, as no one in Argentina was thought to be qualified to do the work (Pyenson 1985).

Bose died in 1911. Three years later another German physicist, Richard Gans, arrived in Argentina and set about the dual task, begun by Bose, of consolidating the "modern" laboratories and training up the students. He directed the La Plata Institute of Physics until 1925, when he returned to Germany to direct the University of Königsberg's Institute of Physics. Under his direction the institute trained Argentina's first doctors in physics, who went abroad to continue their studies, particularly—as to be expected—in Germany, which it is worth remembering was the country with the highest concentration of physicists and cutting-edge researchers in the early twentieth century (c. 1920 almost half the Physics Nobel Prize winners were from that country).

When Gans returned to Germany his successor in La Plata was Ramón Loyarte, the institute's first Argentine director and one of Bose's and Gans's young physics students (Ortiz 2010). Loyarte had worked several stints in Göttingen, Germany, confirming a practice of replacing foreign with local scientists who had acquired experience and training abroad, that is, incorporating "universal" values, topics, and techniques (Kreimer 1998).

"Liberal" Internationalization: Phase One (Early Twentieth Century to the 1960s)

We can talk of the installation of new, local "scientific traditions" from the time the first national leaders of certain scientific fields successfully institutionalized their disciplines. To the earlier examples we should certainly add the birth of biomedical research, embodied in the career of Bernardo Houssay (1947 Nobel Prize for Medicine) who set up the first physiology laboratories at the University of Buenos Aires in 1917. Houssay had devoted himself to developing a number of international networks, particularly in Europe (France and Britain) and the United States, which enabled him to send his students abroad for varying periods and— no less significantly—to assert his international recognition in his own country (Buch 2006; Cueto 1997).[5]

The modality established by the leaders of each disciplinary field in the early twentieth century involved negotiating with the laboratory directors or prestigious institutes with whom local researchers wished to collaborate. Hence, Luis F. Leloir, one of Houssay's most important students, set off for Cambridge, England, in 1936 to "deepen his knowledge of biochemistry in order to understand the intimate nature of physiological processes" (Lorenzano 1994, p. 39). This was clearly a strategic choice on the part of Houssay's laboratory. However, once settled in Cambridge at the laboratory directed by Frederic Hopkins, Leloir gradually began to realign his research towards problems more specifically related to biochemistry. Indeed, on his return to Argentina, although he continued working at the Institute of Physiology, Leloir applied himself decisively to the questions he had already been working on in Cambridge. This trend would be reinforced five years later during his second stay abroad, this time at the Coris' laboratory in Washington University—once again involving a relationship of long standing established by Houssay (Leloir 1982).

We can consider this type of relationship as a true sign of the times: It is characterized by a degree of negotiation between Latin American researchers and their peers from more developed countries. But what was being negotiated? There were three main issues. First and foremost was the topic on which the traveling researcher had to work during his stay, which had to suit both partners. Second were the techniques to be applied and therefore learned by the young researcher, which he would then develop on returning to his country. The third aspect was often to do with the means of financing, as the tools at Argentina's disposal were extremely limited and uncertain.[6] To this we should add an implicit dimension: In most cases, once they were back in their country of origin, "travelers" kept up active collaborative ties with their former "developed" colleagues. These ties were presented in the discourse as the results of the internationalization of science and collaboration among peers, while individual negotiations were based on the principle of a degree of reciprocity in exchanges among peers. This was even more apparent in countries like Argentina or Mexico that demonstrated a degree of precociousness in the institutionalization of scientific research (in comparison with other developing countries, including those of Latin America), to the extent that, like all late-nineteenth- and early-twentieth-century cultural and economic elites, the emerging scientific elite was inclined to put an elevated value on cosmopolitanism.

In terms of the techniques applied, this was a stage—particularly in the life sciences—marked by scant development of instruments and apparatuses, and it was therefore not a prerequisite to have vast resources available. Two examples amply demonstrate the work modalities of the period, marked by the existence of "do-it-yourself" scientists. By the mid-1940s Bernardo Houssay needed dogs so as to extract their hypophysis gland, the function of which he was studying. He made slow progress due to the availability of the animals depending on whatever his students could hunt down. But he had the idea of signing an agreement with the local dog's home, with the institution responsible at the time for rounding up stray dogs off the streets and taking them to specific locations, where most of them were put down. Thanks to this agreement Houssay had a large number of hypophyses available from dogs, and his research took a crucial step forward. A second example: When he set up his own institute (the Campomar Foundation) in 1947, Luis Leloir (another Nobel Prize winner, in 1970) needed a refrigerated centrifuge. He suggested to his students that they make one out of an old washing machine and car inner tubes filled with ice cubes. They then set about making the apparatus, of which Leloir himself was immensely proud (Leloir 1982; Kreimer 2010d).

"Liberal" Internationalization: Phase Two (1960–1970 Approx.)

Two elements played an important role in transforming the noted international relations from the early 1960s: on the one hand, the incipient institutionalization of science policy; on the other, the shift in the nature of research processes in most disciplines. Let us take a brief look at these two aspects.

Argentina's National Scientific and Technical Research Council (CONICET) was created in 1958 and in subsequent years set in motion various public policy instruments, such as the establishment of scientific research degrees and the annual award of grants and subsidies. Even though there was no real thematic gearing of resources granted by the institution at the time, CONICET commissions' deliberations took account of international trends, "hot" topics, and international links as criteria in evaluating proposals. The normative criteria of "good science" (i.e. international science) were thus gradually established implicitly (Feld 2009) through evaluations and the granting of resources.

International links continued to be developed under a regime of laissez-faire, with each researcher or laboratory head endeavoring under their own steam to establish ties with prestigious colleagues from the center, where they were sending their students to work for a period of time or with whom they exchanged information (Kreimer 1998). However the existence of specific financing instruments—previously nonexistent—steered relations towards increasing institutionalization and formalization. Indeed, external grants and travel subsidies began to be awarded systematically, as were research subsidies some time later (Feld 2009). This made it possible to develop more stable and lasting relations, which was accompanied by the presence of specific resources to purchase increasingly important scientific equipment. These changes bring us on to our second aspect: the changing nature of the research processes in most disciplines, brought about by what is generically labeled as the movement towards "big science" (Price 1963; Galison & Hevly 1992).

Indeed, the change in scale and the highly technical nature and substantive increase of the resources needed which came to be present in various fields after World War II brought with them a modification in the more dynamic Latin American countries, rather slower than in the central countries, certainly, but which ended up disrupting the scientific practices that had developed up to that point. Again, two examples, from nuclear physics and molecular biology, amply demonstrate the scope of these changes, which according to some authors were responsible for a shift towards the "fundamentalization" of science at the international level.[7] For historical reasons the orientation towards "basic" science seemed to favor Latin American researchers more, since relations between science and industry have traditionally been weak. That meant an edge—the possibility of developing fundamental research—but also a hurdle: The development of these scientific fields in central countries was accompanied by a close relationship with industry both in their applications and developments—the co-orientation of agendas, that is—as well as the amounts of financing available.

In Argentine physics in the late 1960s, within the National Atomic Energy Commission (CNEA), researchers from the Department of Nuclear Physics began a program on the spectroscopy of gamma radiation using a synchrocyclotron. They set up the conditions to purchase a reliable accelerator, the cost of which exceeded CONICET's budget several times over. Over those years the CNEA developed three nuclear reactors and purchased the first plant to produce nuclear power in Argentina from Siemens in Germany (Hurtado de Mendoza & Vara 2007). A whole parallel community of nuclear physicists and engineers began to be trained around the CNEA, whose ideas and, above all, practices were quite different from what had gone before. Virtually all of this community had spent time training abroad.[8]

In the life sciences the first Latin American molecular biology laboratory was set up in Buenos Aires in 1957, directed by César Milstein from within the Carlos Malbrán Institute of Microbiology. Once this section had been set up, various instruments and equipment were purchased and installed, and Milstein immediately left for Cambridge (United Kingdom) for three years to work with Frederick Sanger. Molecular biology, we should remember, was still an infant discipline at the time, with three central schools of thought (representing three very different approaches): the U.S. school, based in Cold Spring Harbor; the UK school, based in Cambridge; and the French school, based at the Pasteur Institute in Paris (Abir-Am 1992; Cairns et al. 1996). On his return Milstein set himself up as the representative in Argentina of the so-called "structural" school, a continuation of the work of Sanger himself and of Max Perutz. The laboratories were dismantled, however, in 1962 as a result of political intervention, and Milstein left again to settle in Cambridge for good, where in 1984 he obtained the Nobel Prize in Physiology and Medicine for the development of monoclonal antibodies (Kreimer 2010c). The Brazilian tradition of molecular biology, on the other hand, was slightly slower to develop, but was also more solid and very closely linked to the French tradition associated with biochemical research, the so-called "Pasteur" tradition.[9]

One anecdote involving Milstein illustrates the changes over those years in the emergence of a new disciplinary field and the shift from "do-it-yourself" science to the practice of more industrialized research.[10] When Milstein was forced to give up his position at the Institute of Microbiology in 1962, he went to see Leloir and asked him if he could set up at the already highly prestigious Institute of Biochemical Research (Campomar Foundation). But Leloir rejected the idea of creating a molecular biology laboratory and purchasing new equipment for it, arguing that "molecular biology was no more than an auxiliary technique of biochemistry" and that he did not see the "need to acquire such expensive equipment, given that he had developed a highly ambitious research program himself using instruments like a 'homemade' refrigerated centrifuge" (Personal interview with Dr. Milstein, Cambridge 1999, in Kreimer 2010d).

Subordinated Integration in the Liberal
Internationalization Phase

Scientific communities in Latin American countries—like anywhere else, in fact—are not homogeneous spaces of knowledge production. Quite the reverse, they are highly segmented organizations in constant tension. On the one hand, there are effectively integrated researchers taking part in projects and international research programs, regularly attending conferences, handling data that enables them to steer their research in a one direction or another, and often receiving international subsidies. On the other hand, there are groups and researchers who are not well integrated, whose degree of internationalization is poor or nonexistent and whose work takes place in isolation, sometimes gearing themselves towards local needs and often trying to imitate the research agendas of the more integrated groups (Kreimer 2010a).

Such a schematic description aside, the groups that are more integrated in international networks are clearly often the most prestigious within local institutions too. They have the power to determine the orientation at both the level of institutions (policies) and of informal interventions, which influence agendas, the main lines of research, and the most suitable methods. For these researchers there is virtuous circle: Their local "grassroots" prestige enables them to establish links with colleagues from international research centers; their participation in global networks (and external recognition) then decisively grows their local prestige—and power.

I have described the concept of subordinated integration as an important feature of science produced on the periphery. As a direct result of the modality of relating to mainstream scientists, the most integrated groups tend to carry out routine activities: controls, trials, and tests on knowledge already been well established by the teams that take on the coordination in international networks (Kreimer 2000, 2010a). This brings in its wake an important consequence for "peripheral science": Tesearch agendas are often defined within central groups and then adopted by satellite teams as a condition necessary to a complementary style of integration. But such agendas are generally a response to the social, cognitive, and economic interests of the dominant groups and institutions in the most developed countries.

In the long phase of liberal internationalization, lasting until the late 1980s and early 1990s, we are now considering the potential for negotiation was very limited. Yet, in spite of everything, Latin American scientists retained some small room for maneuver, enabling them to influence agendas of collaboration with their colleagues from more prestigious centers and to intervene in the chosen methods and objects of research. The most widespread modality was as follows.

A young Latin American researcher spends a certain amount of time in a central laboratory (thanks to contacts established by his predecessors).

There he or she specializes, say, in mastering a technique chosen by the leaders of each group and applied to an object (e.g. a protein with a specific trait). When he or she returns to their own country, this person usually continues working on the same object and sets himself or herself up as a source of reference at local level, thanks to the technical mastery acquired. At the same time the researcher acts as a supplier of data for the central laboratory that welcomed him or her, as do other researchers belonging to the center itself or to other developing countries. The central group thus exerts cognitive control over the topic in question and—significantly—economic control over possible applications of the knowledge produced. There is an obvious tension in this dynamic: The visibility and scientific quality of local research as legitimized by international groups may contradict the (real and potential) application of the research.[11]

Throughout this stage (and also the preceding one) the disciplinary fields are well established in public research institutions and universities. Against such a background the local leaders of each field see themselves—and operate—as veritable intermediaries between "universal" science and local research. They are the ones in a position to establish lasting ties with international leaders, to whom they send their students on "postdocs", with whom they take part in joint projects, and so on. This strategy enables them to construct the illusion of international integration that conceals its subordinated nature and the tough negotiations they are forced to take if they are to be welcomed into the "global club". In the same operation external recognition enables them to enhance their local prestige: In other words, consolidation of the local position is most often obtained exogenously.

Phase Three: The "new international division of scientific work" and "mega-science"

A number of elements in the dynamics of science and Latin American internationalization would change during and after the 1990s, differing radically from the models we have analyzed previously. These changes are due to a variety reasons.

First, we can see a change in developed countries' S&T policies, which show a significant increase in and concentration of resources aimed at generating "broad regional knowledge blocs" like the European Research Area (ERA).[12] Second, against the background of globalizing trends in recent decades, the massification of communications brought about by electronic media has reinforced the intensity of collaborations between researchers. This modality of collaboration creates an autonomization fiction of the specific contexts where these researchers are based. The situation includes an element of "democratization" in relations governing knowledge production within the framework of 'universalized' ties.

Lastly, it is the very nature of research that has altered, geared as it is towards tackling more complex issues, increasing the number of researchers involved in any one project in correlative terms.

Where European policies are concerned, despite the fact that the official discourse favors the ideology of international cooperation, the instruments applied owe their existence to a strategy of competition with U.S. hegemony in various fields of knowledge.[13] We encounter an even more explicit discourse coming from the United States:

> The rapid changes that have come about at the international level confirm the urgent need to understand and control the place of our nation, its competitiveness, the trends linked in particular to this competitiveness in high technologies, and the critical information that must be generated to better advise the State and the Nation with regard to the future". And also: "Once published, fundamental research may be used freely by all nations, and its results will benefit not only the industries or the countries that financed the research. But [as *Science Board* pointed out in 1993], the advantages for industries and nations that are the first to arrive at discoveries are enormous.
>
> (National Science Board 2008)

Europe and the United States have indeed entered into a global competition over the development of capabilities for scientific research and innovation against the background of a far broader competitive strategy: In response to the enormous bulk of resources the United States devoted to R&D activities through various agencies and with the extremely active role of the private sector, the European Union launched a set of financing initiatives that differed significantly from previous initiatives. For instance, the latest EU Framework Programs have partially displaced calls for projects (largely aimed at the most prestigious scientific groups in European countries) that sought to achieve certain—rather vague— strategic objectives. Instead, a set of initiatives were drawn up that tend to concentrate resources towards a limited number of highly specific networks comprising European institutions, but in which—and this is crucial—the participation of research teams from *developing* countries is strongly encouraged. The scope of the funding increased significantly, and for several years now each network has had opportunities open to it that were unimaginable in the past. The participation of companies in financing R&D projects has also been encouraged, an area in which European countries had always been weak compared to the United States and Japan—with the partial exception of certain research sectors in Germany, the United Kingdom, and Holland.

Let us look at some data that have a great deal to say on the subject: the 4th EU Framework Program (1994–1998) established eleven priority fields (including Information and Communications Technologies, Industrial

Technologies, Environment, Science and Life Technologies, Energy, and Research and Training in the Nuclear Energy Sector) that, given their general scope, covered virtually all fields of knowledge. In addition, each field took in multiple subfields and subtopics. In the life sciences and technologies, for example, a Biotechnology Program (BIOTECH 2) was proposed, consisting of nine broad scientific areas, including cell factories, genome analysis, plant and animal biotechnology, immunology and transdisease vaccination, and structural biology. Each of these areas was further divided into other subtopics and then into more specific subsections. Four hundred and sixty-two projects in total were financed by BIOTECH 2, clearly demonstrating the policy of "broad" and "dispersed" financing mechanisms, with total funding over the four years running to €533 million.

In contrast, the priorities of the 6th Framework Program (2002–2006) were far more clearly focused and, above all, far more specific. The seven priority areas were as follows: life sciences, health genomics and biotechnology, information society technologies, aeronautics and space, food quality and security, global change, and ecosystems. Only the first remains generic in nature, although in its specification and instrumentalization, this is no longer the case. A total of €17,500 million was earmarked for the program as a whole. This trend became more pronounced in the 7th Framework Program (2007–2013), with even more oriented and limited research topics. Table 7.2 shows a brief list of examples clearly contrasting with the "old" forms of presentation:

Table 7.2 Examples of Priority Topics in the 7th Framework Program

- "Upgrading of wood, wood related residues and humus origin substances to added value chemicals and materials: from biological understanding to innovative application"
- "Molecular modelling for rational design of industrial enzymes"
- "New and converging technologies for Precision Livestock Farming in European animal production systems"
- "Development of fermentor-like applications and other plant-based containment systems for molecular farming"

Source: CORDIS (2008)

However, concentrating resources in more focused fields was not the only innovation. Even more important are the new instruments, operations geared to companies, and, above all, the new modalities for project financing: the Networks of Excellence (NoE), the explicit aim of which is to "overcome the fragmentation of European research", and the Integrated Projects designed to "deliver knowledge for new products, processes, services etc." To apply these new instruments, the program provides for the setting up of "consortiums" to concentrate the bulk of resources. The data in Table 7.3 give an idea of the changes at the level of funding.

The changes in the policy, mechanisms, and scope of R&D funding are substantive and, if in the postwar years we spoke of the shift from "little science" to "big science", in recent years we have seen the development of a kind of "mega-science". As we can see in Table 7.3, some research consortia can have more than 1,000 scientists.

Far from being restricted, the active participation of developing countries in these research team networks has been strongly encouraged, *even in official texts* and without them necessarily being teamed up with European teams. In practice, however, it is always European groups who take the initiative in formulating, coordinating, and proposing Networks of Excellence and integrated projects (termed "project leaders"), even when in most networks there is active participation from Latin American research teams. I present an in-depth analysis of Latin American participation in European research—and its consequences—in Chapter 8. However, we can have a quick idea of how the participation of Latin American research groups grew during last years if we see Figure 7.1.

Table 7.3 Financing of Networks of Excellence in the EU

Assistance for integration into NoE	
50 researchers	€1 M/year
100 researchers	€2 M/year
150 researchers	€3 M/year
250 researchers	€4 M/year
500 researchers	€5 M/year
1,000+ researchers	€6 M/year

Source: SCADPlus //SixthFrameProgram (2002–2006).htm

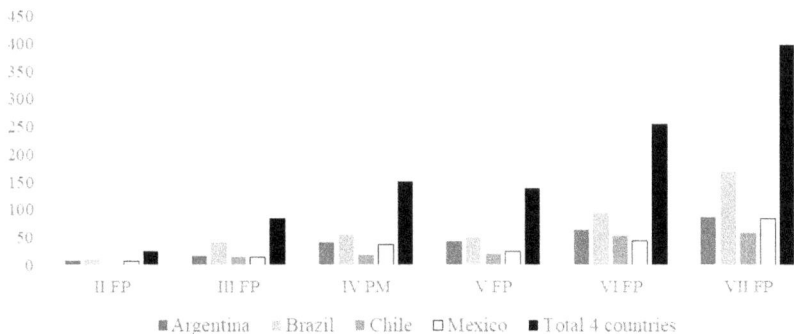

Figure 7.1 Participation (Number of Projects) of Four Latin American Countries in the Framework Programs

Source: Author's own.

2. The Consequences of the New Model for Latin American Research

Given this state of affairs, it is relevant to ask what the consequences of Latin American researchers participating in "mega-networks" are. The traditional modality of "subordinated integration" described earlier has clearly changed in several ways:

- A restriction of negotiating room for "peripheral" teams, who have to integrate themselves in very broad networks and whose agendas have already been solidly structured by the institutions financing them, as well as the public and private actors taking part in them.
- A process of "international division of work" that assigns teams in peripheral countries highly specialized activities with a high technical content but which are subsidiary to already established scientific and/ or industrial problems. In fact, there has been a degree of "relocation" of scientific work, the result of which has been a transfer to the periphery of highly specialized scientific activities that demand high levels of technical qualification but ultimately become routine. In general, all that can be negotiated in these "mega-networks" are subcontracting terms.

When we talk about "established scientific problems", we mean that the research programs have been conceived by the leaders of the hegemonic groups—regarding both conceptual issues and the methods and instruments to be used—and it is only "after the fact" that peripheral researchers are invited to take part. This restriction is reinforced in the case of industrial/scientific projects: Researchers in these situations have already negotiated with the companies belonging to the project, and there is usually no chance for peripheral researchers to draw attention to their own cognitive interests.

- Research teams from the periphery participating in "mega-networks" have the chance to significantly increase their resources and integration connections, while their reproduction is amplified by the fact that these teams incorporate new researchers who are trained in this way. Their stays at international centers of excellence are strictly functional the new dynamics: They consist of periods spent learning new methods and techniques that they will later apply on returning to their countries. Not just anyone can be the subject (or object) of subcontracting, as a certain level of excellence and recognition by peers from the international community is required.

The three characteristics of the new model lead us to believe that the most serious tension generated in this framework is bound up with the local relevance of the research, that is, its social utility for the community where it is conducted. Indeed, this new kind of internationalization leaves a very little room to formulate social and local problems as knowledge problems.[14]

The process of change can be analyzed on two levels. At the formal level, while in "liberal internationalization" a greater degree of freedom was enjoyed by local teams, the justification of local research agendas in terms of social or economic needs was at odds with the researchers' international ties, but the two approaches did not appear to be mutually exclusive. The explicit aim of local researchers was to produce "superior-quality" knowledge, and their research was justified by the general progress in knowledge, an idea founded on a collective belief—held in particular by science policy-making bodies—in the linear model of innovation. According to this model the creation of significant stocks of (fundamental or applied) knowledge would act as an engine to move the heavy cogs and would end up bringing useful innovations to all social actors. However, on another level of analysis, the effects of this model are more symbolic than material: Most knowledge produced under this logic was of more use in increasing local researchers' visibility than creating knowledge that was useful and applicable locally.

Defining the social needs that may be subject to a "demand for knowledge" is a far-from-simple matter: It involves asking questions about which actors have the legitimacy and capability to formulate such demands.[15] This also involves determining the mechanisms through which social problems are translated into knowledge problems. This is particularly important, since the actors experiencing the most pressing social needs are precisely those that have most trouble performing the operation of translation. There is usually a group of spokespersons who represent many others that have no voice. Two of these spokespersons are particularly important: the scientists themselves and, through various agencies, the state.

3. Two Cases Illustrating the New International Division of Scientific Work[16]

Seagulls and Whales in Patagonia

The Puerto Madryn region is one of the main tourist centers in Argentine Patagonia due largely to whale watching. The whales arrive each year between September and January to mate and give birth. This event—which has cultural and economic consequences—has also been the subject of scientific research for several decades, in particular, at the National Patagonian Centre (CENPAT), a CONICET-run institution.[17]

When it comes to formulating social problems vis-à-vis scientific problems, it is interesting that an "attack behaviour by kelp gulls (*Larus dominicanus*) towards southern right whales (*Eubalaena australis*) in the

184 Centers and Peripheries Revisited

waters off the Valdés Peninsula was first recorded in the early 1970s" (Sironi, Rowntree, Snowdon, Valenzuela, & Marón 2005). It is worth noting that biologists and ecologists used to talk about whales as a "charismatic species", even though they have a somewhat intuitive conception of what "charisma" really means: As Albert, Luque, and Courchamp (2018) point out, "charisma is a term commonly used in conservation biology to describe species. However, (. . .) the term 'charismatic species' has never been properly defined". Nevertheless, we can take the term in its native sense and highlight how a social, political, or ultimately human characteristic has been transferred to the animal world.

This kind of gulls' behavior had never been seen anywhere else in the world or with any other species of whale. From then on the gulls' attacks on whales spread from the San José Gulf to the entire Valdés Peninsula, rising more than fivefold between 1985 and 1990. The seagulls feed on pieces of skin and blubber from exposed areas of the whales, producing wounds that increase in number and size in each individual during the season, which runs from July to December.

Juan, a CENPAT biologist, went to do his postdoctoral studies in Kiel and Bremen, Germany, in 1985 in order to approach this subject.[18] On his return to Argentina he consolidated his position in CENPAT by strengthening his ties with his German colleagues. "The main researcher, a couple of European scholarship holders and the odd technician or support worker" arrived in Patagonia from Kiel and Bremen to collect data on seabirds using electronic tags to carry out different measurements.

The unit cost of these devices was around US$2000, which it was impossible to finance using local resources, particularly because many of these sensors would be lost during the fieldwork (fourteen devices were used for the initial part of the project). The data collected served as material for the scholarship holders' doctoral theses, while the local researcher was given the opportunity to codirect two theses and, most notably, to publish coauthored articles in "extremely high-impact" international journals, albeit always signing as the second author. In addition to the local leader three other researchers traveled to Germany on various training trips. The disparity of resources was huge: The German center had at its disposal seven oceanographic vessels and even used techniques from the nuclear physics group at the same institute.

The whales' destinations were previously unknown outside of this spell in the Puerto Madryn region, but with these sensors it was possible to monitor them all year round. However, the data provided by the sensors were processed in the laboratories of the University of Bremen, the Research Center Ocean Margins (RCOM), and the University of Kiel through an international network called Cluster of Excellence "The Future Ocean". Here data are processed from the various oceans around the world. This meant that the researchers from Patagonia became "privileged producers" of information that would later be globalized.

A good deal of significant information was produced and published in various journals coauthored by Argentine and German researchers. However, despite the local researchers' best efforts to reverse the gulls' effect on the whales, the situation has not improved: Possessing only fragmentary data, they have not been able to use the information to tackle the problem at a local level. The percentage of whales subject to attack by gulls in 1995 was 26 percent; by 2008 the figure had risen to 90 percent (White, Gillon, Black, & Reid 2002; Bertelotti & Pérez Martínez 2008).

Plasma Physics on the "Periphery of the Periphery"

In 1981, two young researchers who had completed their degrees in physics at the University of Buenos Aires (UBA) and were working at the UBA's Institute of Plasma Physics (INFIP) moved to Germany to spend two years at the University of Düsseldorf. The first of the two went to do postdoctoral research after finishing his doctorate; his fellow researcher went there to complete her doctorate. A few years earlier, in 1976, they had both spent time in the Institute of Nuclear Physics at the Frascati National Laboratories in Italy with funding from the Italian government to receive training in the use of time-of-flight measurement techniques for dense plasma focus. The contact for this activity was INFIP's director, who had worked for some years at the Italian institute, where he had got to know Dr. Konrad, a researcher from the University of Dusseldorf and a guest lecturer at the Frascati Laboratories (Taborga 2010).

Through Dr. Konrad that they were put in contact with Dr. Decker of the University of Düsserdof, who took an interest in them while directing a project to create a high-tension plasma focus device. They were therefore given the task of studying the temperature of the plasma they produced in their plasma focus in order to ultimately improve the device by trying to "scale up", to have more neutrons from which to obtain energy.

Dr. Decker allowed them to take part in the design, and they were able to apply "their own criterion about neutrons": They studied temperature using neutron time-of-flight spectroscopy to calculate the density of the plasma in relation to the voltage. This enabled them to observe— "discover" in their account—that the plasma was being formed inefficiently: The gas was not ionizing completely, due to problems in the equipment's design.

Back in Argentina, instead of returning to the UBA, they set themselves up in Tandil (400 km from Buenos Aires) at the Arroyo Seco Institute of Physics (IFAS) of the National University of the Centre of Buenos Aires Province, a medium-sized and relatively new institution compared to the country's "great research traditions". Most of the institute's lines of research usually replicated the work of the UBA's Institute of Physics, with whom historically they had a relationship of collaboration/dependency. When the two researchers returned, they suggested opening up a new line of research and manufacturing a plasma focus in Tandil. The project was conducted between the late 1980s and the early 1990s in close collaboration with and, significantly, with financing from Dr. Decker's group.

As a result, two German researchers arrived in Tandil, together with two engineers, to coordinate the manufacture of the plasma device, a process that took over two years' work. Once it was operating, they coordinated the work to be undertaken over the next three years with the German director. This basically involved complex measurements that took up most of the time. In exchange the Argentine group was able to participate in a dozen publications in tandem with their Düsseldorf colleagues, all of which emerged from taking measurements with the plasma focus in Tandil. It was almost impossible for them, therefore, to develop autonomous lines of research, independent of the "agreements" they had with the German group. In the third year the director and two scholarship holders began working at the institute at the weekend to "get in on the act" and to get to use the apparatus for their own projects. The work "in secret" (to use the researchers' own words) continued until the mid-1990s, when the new director of the German group declared the Tandil plasma focus obsolete and there was no point in continuing to use it for the measurements.

4. A New Kind of International Scientific Integration for Latin American Countries

In the course of this chapter I have tried to show the development of science in Argentina through the role of relations between local researchers and their peers from the central countries. The hypothesis underlying this periodization is that, far from playing a secondary role in the strategies of research groups as part of the inevitably universal nature of science, international ties are fundamental in understanding the organization of local scientific traditions and their historical developments.

So, throughout the early decades of the twentieth century, international ties were organized along the lines of the institutionalization of various disciplinary fields in a firm bond with European scientists (and some from the United States), making the emergence of local leaders possible. During this period the equipment used—and hence the technical research decisions—were still defined by ties with scientific centers in the mother country. During the following period local leaders began to implement their own strategies and use their international connections to a threefold end: (a) to align themselves with international agendas, thereby proving the "modern" nature of their lines of work; (b) to obtain recognition from their foreign peers (with whom they even published papers), which they would then assert before their colleagues in the local context; and (c) to send their students abroad to train or fine-tune their skills in prestigious laboratories, enabling them to reproduce and rein-force local research traditions.

These strategies became consolidated towards the end of the 1950s, when science and technology policies were institutionalized in Argentina. The nature of the connections was not radically modified, but was reinforced with the support of precise policy instruments, such as the scholarship system that enabled young researchers to be sent abroad. At the same time, local policies, particularly the availability of a system of research grants, allowed equipment to be modified in a scientific world that was becoming increasingly complex in terms of technical research requirements, higher costs, and the changing scale of research equipment (this first became apparent in disciplines like physics rather than the bio-medical sciences, whose industrialization came a little later). The strategy of "do-it-yourself" scientists like Houssay and Leloir manufacturing their own apparatus was no longer conceivable: Access had to be gained to an international research equipment market if one was to start working with standardized apparatus, which had to be described in detail in articles. To do this, local leaders had to seek resources abroad and embark on fresh negotiations.

By the 1990s the context has altered profoundly, with the emergence of new policies in developed countries and the change in the structure and scale of research, giving rise to what I term "mega-networks". In the three cases chosen to illustrate this new stage there was considerable centralization in the cognitive control of research topics, including the definition of methodological and technical questions, which were by now highly structured. In both plasma physics and the Trypanosoma Cruzi Genome Project the margin of autonomous decision-making open to local researchers where their research plans were concerned was mini-mal, as they became qualified data producers. This task should not be downplayed as a mere "technical application," as integration into these networks entails high scientific standards and highly developed techni-cal capabilities in order to interact with the international leaders of each mega-network. This enables local leaders to increase their scientific out-put in international journals, often coauthored with international col-leagues. But the most notable consequence of this dynamic is twofold: (a) the ability to determine lines of research autonomously (a process accompanied by local S&T policies, which welcome any international collaboration irrespective of the contents involved) and, above all, (b) in terms of the possibilities of local industrialization of locally produced knowledge.

This dynamic is not the only one to emerge in the context of Latin American research, of course. Other strategies also coexist, such as the attempts by certain groups—usually less internationally integrated and hence less prestigious—to steer their research agendas autonomously. This is, for example, true of groups dedicated to public production of medicines whose patents already belong to the public domain. However, public policies are contradictory in that the public discourses are geared towards the production of knowledge for local use and to tackle social problems (Kreimer & Zabala 2009), while the instruments actually implemented are geared more towards a logic of international cooperation based on an "ideology of intensity". In other words, these are policies that favor the intensity of international cooperation, regardless of the nature and content of such ties. This has been reinforced in Argentina—in contrast to other Latin American countries such as Mexico, Colombia, and, above all, Brazil—by a shortage of scholarships for training and research abroad over the last twenty years at least, and their virtual absence for the last ten years. In practical terms there is no domestic training policy for studying abroad, but it is strongly dependent on the offer of institutions located in developed countries (or international agencies). If in the past this dynamic contributed to the "brain drain", by exogenously determining stays abroad, it nowadays weakens local scientists' chances of negotiating better research (and international integration) strategies.

Notes

1. "Liberal" is used here to mean practices not regulated by national authorities or an institution's management. Rather, it refers to practices characterized by a *laissez-faire* approach that only obeys the strategies of the researchers themselves.
2. The expression "oriented liberal" may seem contradictory. However, as we see later on, I am referring to the startup of international cooperation support mechanisms that nevertheless did not affect researchers' freedom to establish international ties of their own accord.
3. This new dynamic of the international division of scientific work into large networks is certainly not the only one we can observe. Traditional relations continue to be developed, as do more complex modalities. However, we may look on this emerging dynamic as eloquently marking and prefiguring the tensions we will see unfold in various scientific fields in the years to come.
4. While certain works have quite rightly postulated that it is inadvisable to ignore the scientific dynamic that became apparent following independence (c. 1810) with the introduction of positivism (Saldaña, 1992, 1996), I take the last quarter of the nineteenth century as a turning point, as it was during this period that a real and lasting process of institutionalization in "modern" Latin American science got underway.
5. For information on the role of scientific elites see the classic works by Whitley (1985) and Mulkay (1976). For the role of scientific elites in Latin America see Vessuri (1984, 1996c) and Kreimer (2006).
6. However, most of those who gained access to research training in the 1940s came from wealthy families who helped cover their foreign travel and accommodation expenses. (Leloir's family, who were important landowning *estancieros* (farmers), is a typical case, but there were many others; e.g. Braun Menéndez, Castex, et al.).
7. Indeed, Pestre (2003a) points out that this process was accompanied on the international plane by a new "fundamentalization": on one hand, a material ability to express and manipulate phenomena at the level of elemental

entities (the atomic nucleus in the physical sciences) or molecular entities (in biochemistry and biology), and on the other, an ability to first measure and purify and then recompose and utilize those entities (e.g. in producing "molecular jets" or sequencing genes).

8. For example, one of the most conspicuous researchers of the period, Juan José Giambiagi, had completed his postdoctoral studies in Manchester, Daniel Bes in Copenhagen, Mario Mariscotti had worked at the Brookhaven National Laboratory, and Edgardo Valenzuela had done his doctorate at Stanford.

9. In Argentina, the English tradition through Milstein and the French tradition through Ignacio Pirosky, then director of the Malbrán Institute, were present at the institutionalization of molecular biology. The third, known as "informational", was represented by Delbrück, Luria, and Hershey, linked to the famous "Phage Group" in the United States (Cairns et al. 1996, Mullins, 1972). See Chapter 5 for the full story of molecular biology in this country.

10. On the subject of the life sciences, Gaudillière (2001) notes for example that, in the United States, the breeding of mice for laboratory experiments took on an industrial quality and rose steadily from the 1940s. According to Gaudillière, mice became veritable "instruments" provided by the industry to researchers.

11. This idea is developed in Kreimer (2006). The situation was first brought into sharp focus by Varsavsky (1969).

12. One of the explicit goals of the ERA is to "develop strong links with partners around the world so that Europe benefits from the worldwide progress of knowledge, contributes to global development and takes a leading role in international initiatives to solve global issues" (European Commission, 2007b).

13. The report states that "Science knows no boundaries and the issues that research is asked to deal with are increasingly global. The challenge is to make sure that international S&T cooperation contributes effectively to stability, security and prosperity in the world" (European Commission, 2007b, p. 23).

14. For an analysis of the relationships between social and scientific problems, see Chapter 3.

15. On this topic, see Bourdieu (1997) who has a radically critical view of the concept of "social demand" of knowledge, and Gusfield (1981), who had—in my opinion—the finest analysis on the emergence of public problems and the role played by the scientific knowledge.

16. I am grateful to Manuel González and Ana Taborga for their help on the case studies.

17. CENPAT was set up in 1970 as an interdisciplinary center for research into various aspects of regional interest. It began reporting to the National Scientific and Technical Research Council (CONICET) in 1978.

18. We have agreed to respect the researchers' anonymity; therefore, "Juan" and other names used here are fictitious.

8 Globalization or Neo-Imperialism?

Latin American Science in the Era of Globalization

In the previous chapters I analyzed the local and international dimensions affecting both the production and social use of scientific knowledge and the formulation of social and scientific problems. We have seen also how specific scientific fields emerged and grew along several decades in the heat of these tensions. Finally, we observed the various historical stages of the internationalization of Latin American science. I will analyze in this chapter these issues from another angle related to international scientific cooperation, in this case, the participation of Latin American scientists in European projects.

The chapter is based on research conducted in recent years, where we analyzed the participation of Latin American scientists in consortia funded by European Framework Programs (FP), especially the 6th FP (from 2002 to 2006) and the 7th (from 2007 to 2013).

The approach adopted here is certainly original, since we observe both points of view: those coming from Latin American researchers and those coming from the European leaders of some specific research consortia. In fact, during the study we collected data, through different methods, about the opinions and views on both sides of the Atlantic Sea.[1]

Therefore, after discussing the general aspects and providing the reader with contextual elements necessary to understand these issues, I will first present the views of Latin American leaders who participated in European projects and then the perspective of the European coordinators. In a nutshell, there are two types of questions inspiring our research: first, why European leaders enlist Latin Americans to work in their consortia, and reciprocally why Latin American scientists are happy to take part in these projects. Second, what kind of activities do Latin Americans perform when they participate in European research consortia?

To a large extent, the research was inspired by previous inquiries (presented in Chapter 7), which gave us evidence of the emergence of a new international division of scientific work. However, in previous works we had observed some general features and corroborated these findings in certain particular cases analyzed at the micro level: the conflictive relationships between whales and seagulls, the construction of a plasma focus device, and the production of drugs and vaccines for Chagas disease, among others. However, broadening the inquiry to a greater number of cases, even if this implies—as it always happens in the social sciences— losing the details of each case in particular, allowed me to strengthen the general hypothesis and even observe other facts that had not been evident until then.

1. The International Scientific Cooperation

It is a well-documented fact that cooperation among scientists from different countries as measured in publications has steadily increased, particularly over the last three decades (Adams 2012, 2013). Leydes- dorf and Wagner (2008) point out that "[i]nternational collaboration as measured by co-authorship relations on refereed papers grew linearly from 1990 to 2005 in terms of the number of papers, but exponen- tially in terms of the number of international addresses", confirming the hypothesis of Persson, Glänzel, and Danell (2004) about an infla- tion in international collaboration. The proportion of internationally coauthored rather than "home-grown" articles is growing in a signifi- cant group of countries (in the United Kingdom and Switzerland, it is already over half), while it is impossible in some small countries to distinguish "domestic science" (Adams 2013). Whereas in 1988 just over 10 percent of papers were signed by researchers from more than one country, twenty years on, that figure had increased to 30 percent (Boekholt, Edler, Cunningham, & Flanagan 2009; Gaillard, Gaillard, & Arvanitis 2010). Against such a background it is not surprising to find that Latin American scientists' cooperation on projects undertaken with groups and colleagues from more developed countries has risen very significantly over this period (Gaillard & Arvanitis 2013; Kreimer & Levin 2013).

The definitions, scope, actors, institutions, regulations, and specific practices of international scientific cooperation have been the object of numerous ambiguities and debates. Therefore, before delving into the specific analysis of our object, at least two points need to be established relating to North–South cooperation and linked with international scientific cooperation in order to set the parameters of our research object.[2]

North–South cooperation presents specific interpretative problems and tensions. This is due first and foremost to the definition of the actors in play itself: Some works on North–South cooperation refer to "devel- oping countries" as a relatively homogeneous group. Wagner (2008), for example, points out that, after the emergence of new networks as a means to organize twenty-first-century knowledge production, there are several operations in the uses of knowledge that remain in the local sphere, and developing countries therefore need to establish and develop a series of institutional arrangements if they wish to take advantage of these opportunities. According to Wagner (2008, p. 115):

> Indeed, developing countries have an advantage over developed countries—they have not built to twentieth-century national science system. This may seem counterintuitive; after all, most developing countries want to have highly developed scientific capabilities. But because these countries do not have the embedded twentieth-century bureaucracies and institutions that were the hallmarks of the era of scientific nationalism, they have greater flexibility to pursue new developments in science. The absence of nationally driven constraints tied to a huge investment can actually be an advantage that developing countries can exploit by building a more nimble networked system.

This optimistic version, which is certainly seductive, has a number of drawbacks, the first of which lies in declaring the extinction of the nation state as a space for science: In parallel to networks, there continue to be national policies, local applications of knowledge, and knowledge strongly linked to its localized spaces of production and use (Arvanitis 2011).

Another problem is that the label "developing countries" includes only those with limited scientific traditions or weak national science systems. However, countries like Argentina, Chile, Brazil, Mexico, Egypt, South Africa, and so on that belong to the "developing world" at large have been highly dynamic in terms of knowledge production and scientific systems for more than a century. Gaillard et al. (2010) have shown that the performance of international cooperation in Latin America differs widely according to individual countries' scientific robustness. Even if between 1985 and 2006 a steady rise in international copublications is seen in the region at the expense of publications by same nationality authors, in countries like Brazil, Argentina, Mexico, and Chile with larger scientific communities and with higher scientific production, the share of international copublications is lower than in countries with less developed scientific systems.[3] That's why I have chosen Latin American countries with similar levels of socioeconomic and scientific development, enabling more solid inferences to be drawn.

Second, North–South scientific cooperation is often couched in a dual discourse by both European and Latin American decision-makers as something "good in itself": On the Latin American side, it is something impregnated with cosmopolitanism, the benefits of interacting with world leaders and greater visibility for local science. On the European Union's side the discourses are more pragmatic, referring to cooperation as a strategic resource for strengthening European science and, above all, competitiveness.

The relevant literature, however, highlights certain tensions, in particular "optimistic" versus critical views. On the one hand, some studies have echoed the optimistic outlook, celebrating the intensification of cooperation as a means to develop the cosmopolitanism of researchers from the South and integrate them in "international science" (Sebastian 2007) or underlining the democratizing effect of the new forms of international cooperation and the "opportunities" that present themselves for developing countries (Wagner 2008; Anderson 2011). On the other hand, more critical perspectives have emphasized the asymmetries, subordinated relations, or even dependency that structures the modes of collaboration among more or less developed contexts (Gaillard 1994; Vessuri 1996b; Cetto & Vessuri 2005; Velho 2002; Kreimer 1998; Kreimer & Meyer 2008; Kreimer & Levin 2013; Beigel 2014). At stake is whether such cooperation improves developing countries' cognitive and technical capacities, whether it is a mode of marginal and subordinate insertion, whether there are asymmetries in the setting of research agendas, and whether these agendas are suited to developing countries' needs. This chapter explores these questions using a methodological strategy focused on work dynamics within research consortia in order to chart the complex, often contradictory nature of relations of international cooperation among groups with different levels of development.

Where broader international scientific cooperation is concerned, Katz and Martin (1997) have pointed out the ambiguous definition of what is understood by cooperation. For example, there is an ample body of literature that takes scientific publications as material for the study of relations of international cooperation and coauthorships (the bibliometric perspective) on the assumption they are a good proxy of these practices (Gläser & Laudel 2001; Newman 2001; Leydesdorff & Wagner 2008; Wagner & Kit Wong 2012). While this macro-methodology does bring out a certain dynamic in terms of major trends in cooperation in given countries, in specific disciplinary fields and even in some thematic clusters, it shows only a part of the product of collaborations (papers) and tells us nothing about the methods of organization or motivations behind cooperation, despite these being crucial to an understanding of scientific relations between centers and peripheries. Indeed, many social activities and practices do not necessarily crystallize in scientific articles, but instead in many other ways, such as exchanges of researchers and fellows, the development of joint training programs, industrial product development, the organization of seminars, and so forth.

Considering the factors underlying the increase in international collaborations Wagner, Brahmakulam, Jackson, Wong, and Yoda (2001) suggest the main reasons for enhanced international cooperation: geographic proximity, history, common language, specific problems and issues (common problems, such as disease control or natural disaster mitigation), economic factors, expertise ("Many developing countries have institutions and individuals with world-class expertise . . ."), research equipment, databases, and laboratories.

Some of these possible reasons may seem original and explanatory, albeit not empirical. For example, geographical proximity is not a common "strong" reason for collaborations; in Latin America, as well as in other regions (e.g. South Africa), research groups are more frequently associated with groups from "the North" than with those akin or geographically close to them. The "common language" does not seem to be decisive either. Many studies indicate that "specific problems and issues" can be quite challenging since the work agendas are often strongly influenced by the more advanced countries (Bradley 2007; Gaillard 1994; Kreimer & Meyer 2008; Vessuri 1996b).

Other causes seem more plausible, such as economic factors or research equipment, databases, and laboratories, and can better explain the motivation of the groups located in nonhegemonic contexts, on the condition that their access to resources is, indeed, seriously limited.

In general terms, we can identify three contextual elements to explain some of the changes in the organization of cooperation in recent decades. The first involves the changes in scale beginning in the late twentieth century, when there was a shift from so-called "big science" (Price 1963; Galison & Hevly 1992) as it unfolded in the postwar period and throughout the Cold War (Hallonsten 2016) to science on a far larger scale, one of whose expressions—though not the only one—is various "mega-projects" (Beaver 2001; Wagner 2008; Kreimer 2012), such as the sequencing of the human genome or the Large Hadron Collider, mobilizing thousands of researchers. Unlike the traditional model of big science, which has been associated with a brain drain from developing countries (Devan & Tewari 2001; OECD 2002), these projects do not necessarily involve the settlement of many researchers in the same physical place; with the spread of ICTs scientists from a variety of geographical locations can take part in the form of networks (Shrum 2005; Adams 2012).

The second involves the reorganization of traditional disciplines and the emergence of new fields. Areas of research have effectively become more complex, generating displacements and hybridizations across disciplines—and across academic and industrial research—in "technological research communities" (Joerges & Shinn 2001) and new regimes of knowledge production (Pestre 2003b).

The third involves policies deployed by developed countries designed to stimulate international cooperation, which have imparted explicit momentum to such linkages both in discourses and the implementation of specific instruments: The European Union (European Commission 2008) and the United States (National Science Board 2008; National Science Foundation 2008) have had specific instruments in place to promote international cooperation for several decades. Europe has promoted it at both the aggregate and national level, while the United States has done so in a diversified way through numerous public and private institutions and agencies (Whitley 2010).

These three elements are important to an understanding of the specific type of cooperation under discussion here: one resulting from explicit European Union policies and directed at predetermined targets in each thematic area's programs and in calls for projects. This involves the formation of medium or large international consortia lasting approximately four years in which each group plays a specific role within the topic or issue in question, transcending disciplinary barriers.

Given that European Union policies play a fundamental role in the kind of cooperation analyzed here, a distinction needs to be made—as much of the literature has—between the political and individual motivations behind cooperation and how these join up with other factors that stimulate or obstruct it (Beaver 2001; Bozeman & Corley 2004; Boekholt et al. 2009; Edler & Flanagan 2011; Wagner 2006, 2008; Gaillard & Arvanitis 2013). In the next section we therefore look at some of the general characteristics of extra-European cooperation policy within the Framework Programs to gain a better understanding of Latin American researchers' roles in international research consortia.

North–South Cooperation and EU International Cooperation Policy

Taking into consideration the scientific cooperation policies of international organizations and funding agencies located in developed countries, Gaillard (1999) identifies three successive phases in North–South scientific cooperation. The first runs from the colonial period through to the 1960s and 1970s and is focused on finding solutions to the problem of development through the mobilization of (human and financial) scientific resources from the countries of the North. The second unfolded between the 1970s and the 1980s/1990s and focused on endogenous capacity building in the countries of the South. The third and most recent is geared to producing cooperation structures with the leitmotif of mutual benefit.

The European Union's policies of scientific cooperation with developing countries have followed criteria similar to those described by Gaillard. Such policies can effectively be traced back to the 1980s, when the European Parliament set up the Science and Technology Development Program (STD). This had three phases: STD I (1983–1987), STD II (1987–1990), and STD III (1991–1994). These programs aimed at strengthening research capacities and increasing the impact of research in developing countries, especially in fields such as tropical and subtropical agriculture, on the one hand, and medicine, health and nutrition, on the other (Gaillard 1994).

After the Fourth Framework Program (FP4) (1994–1998) the European Union created a special subprogram to foster "Co-operation with

Third Countries and International Organizations" (INCO), which included cooperation with developing countries (INCO-DC). INCO-DC expanded the range of areas, including some not related strictly to developing countries' needs: (a) renewable natural resource management (forests, oceans, water, energy); (b) agriculture and agro-industry (improved production, storage and marketing); (c) health (disease control, vaccines, support systems); and (d) topics of mutual interest established by agreement in sectors like information and communication technologies, new materials, and so on (Gusmão 2000).

After the Sixth and particularly the Seventh Framework Programs (FP6 and FP7) there was an interesting new development: cooperation with developing countries presents no significant differences in terms of the policy instruments to intra-European or cooperation with other developed countries (European Commission 2005). To understand this innovation a very brief outline of the four subprograms of FP7 (very similar to FP6) is helpful (European Commission 2007a):

1. *Cooperation*: Funding for projects of European-backed international consortia and third countries in the following priority areas: health, agriculture, fisheries, and livestock; information and communication technologies; nanosciences, nanotechnologies, and materials; energy, environment, transport, socioeconomic, and human sciences; and space and security.
2. *Ideas*: Funding for research excellence in individual groups not necessarily comprising third countries.
3. *People*: Funding for intra- and extra-European mobility projects.
4. *Capacity Building*: Provision of four instruments in which third countries can participate: infrastructure for research, research to benefit small and medium-sized enterprises, regions of knowledge, research potential, science in society, and specific international cooperation activities (INCO).

This structure no longer channels third-country cooperation exclusively through the INCO subprogram (which becomes a funding line aimed at networking activities and building biregional agendas), and third countries can also participate on an equal footing with the member states of the "cooperation" subprogram. This is the most relevant in budgetary terms, absorbing 69.5 percent of the FP6's resources and 64.1 percent of FP7's (https://cordis.europa.eu/fp6/budget.htm and European Commission, 2007a).

Behind this organizational change lies a vigorous policy of strengthening for the European Research Area (ERA), to which end deepening intra- and extra-European international relations is seen as a necessity. This policy was expressed in a 40 percent increase in FP7's annual budget compared to FP6, an upwards trend that continued in the recent Horizon

2020 (Muldur et al. 2006). As we have seen in Chapter 7, this largely accounts for the increase in Latin American participation: research groups from the region participated in 308 projects in FP7 (2007–2013), compared with 204 projects in FP6 (2002–2006) (CORDIS 2009).[4]

It is as well then to ask what motivations underlie this policy. In an attempt to summarize and classify the motivations behind international cooperation policies, Boekholt et al. (2009) distinguish the "narrow paradigm" from the "broad paradigm": The first one relates to objectives within "science policies", how to improve the quality, scope, and critical mass of research or *to make the European research area more attractive to highly qualified human resources from third countries*; the second one refers to aims within "policies through science" like improving competitiveness, confronting global social challenges, and supporting least developed countries in science and technology capacity building. In either case it is a strategy to overcome the European region's disadvantages compared to other global competitors like the United States or Japan.[5]

On the subject of the "broad paradigm" a European Union document (2008) stated the following:

> Europe can play a more active role in international agenda setting and formulation of policies and strategies and be more audible in international negotiations. Taking the fore on the international policymaking scene can be a way for the EU to reinforce the bases of its economic competitiveness in the future through influencing early the design of international regulations affecting its private sector.
>
> (European Union 2008, p. 29)

This concern over global agendas and competitiveness has led to two trends in the Framework Programs. The first is a higher concentration of resources for better-defined and more restricted (realistic) goals in calls for projects (Kreimer 2006). The second is an increase in company participation: If there was a fall in companies' participation in research consortia between FP4 and FP6, the EU's economic contribution to industrial companies rose from 17 percent in FP6 to 25 percent in FP7, while the percentage of contracts with industrial companies compared to other participant institutions rose from 19 percent to 30 percent between the two programs (European Commission 2016). A 48 percent rise in the funding contribution was concentrated in the various thematic areas comprising the subprogram "cooperation", with 21 percent due to the introduction of new instruments from FP7 (European Commission 2016).

Consequently, despite the fact that the Framework Programs include financing instruments for projects to target global challenges (like climate change) or to boost research capacities and solve problems in developing countries (like endemic diseases), the leitmotiv of North–South cooperation of the 1970s–1980s has been losing ground to a policy which in the guise of "free access" to European instruments and financing and "mutual benefit" poses a fresh challenges regarding the (real or potential) asymmetries involved in the setting of the agendas and the (private) exploitation of knowledge.[6]

2. Latin American Views: Motivations, Strategies, and Cooperation Results

I summarize the qualitative features of the scientific cooperation activities of Latin American groups that obtained funding in projects covered by FP6 and FP7. These findings are based on a survey we conducted among regional leaders during 2015.

Start of and Motivations for Participation

The first issue relates to whether participation in European networks is an important factor in orienting the lines of research followed by Latin American groups. The results obtained suggest that participation in European projects has virtually no influence on generating new lines of research, but, on the contrary, Latin American groups participate on the basis of pre-established areas of activity. Sixty percent thus state that the project "falls within their group's traditional lines of work", while 30 percent indicate that "it focuses on topics akin to their group, even if it is not their main line of work". Therefore, participation in EU projects seems not to be a main driver for thematic innovation: only 10 percent, therefore, acknowledge having embarked on new topics as a result of their participation in the European network.

This apparent thematic neutrality may respond to another reason: Insofar who participate in European projects are Latin American elites, these groups have historically oriented their agendas following dominant lines in hegemonic research centers. As a consequence, they are enlisted by European scientific leaders precisely because of their previous alignment.

With regard to the commencement of their involvement in European projects and their motivations for participating, a very large proportion of respondents indicated that a key factor was the invitation from a European researcher, be it the project leader (48 percent) or another European group (10 percent) (see Figure 8.1). Prior relationships between the groups (17 percent) and informal conversations (8 percent) also carried a degree of weight. As we can see, institutional initiatives, in particular international cooperation policies, do not seem to play a vital role here from the researchers' point of view, which is quite striking, given the strong presence of specialized offices in the leading countries (Argentina, Brazil, Mexico, and Chile) to encourage participation in European projects. Indeed, only 5 percent of the respondents replied that these actions were influential in the commencement of their participation.

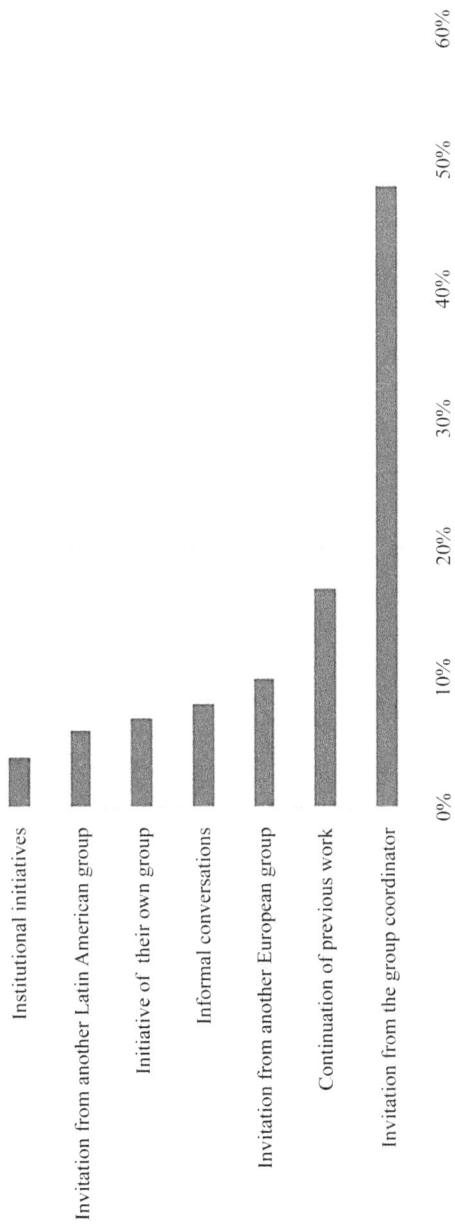

Figure 8.1 Starting Participation in European Projects (%)

The motivations given by the Latin American leaders for partici- pating in European projects show that the most important reason lies in "an interest in the subject area", since almost 60 percent ranked this first (RK1), which is consistent with the fact that the vast major- ity participate in networks that involve a continuation of their lines of work (see Table 8.1). We believe that the remainder of the responses were influenced by a sense of obligation to provide a particular answer, rather than their true motivations: For example, as a first option, "the possibility of linking up with prestigious groups" is mentioned later, which is not surprising, but is no doubt underestimated by the research- ers themselves. On the other hand, "access to funds" and "access to equipment" are the motivations most frequently mentioned as a second option (RK2, with 17.7 percent and 21 percent, respectively), which leads us to think that, for almost 40 percent of the researchers, this is an important motivation, even though it is not the first one they mention spontaneously.

Degree of Freedom to Define Topics, Methods, and Organization of Study

It is interesting to explore the participation of Latin American leaders when it comes to defining the research subjects and methods employed, since a prior hypothesis suggests that these topics are already firmly established when they are invited to participate. The results partially confirm this hypothesis, but with some qualifications: In terms of the research topics, almost two-thirds (64 percent) declared that these were already either "clearly established" (33 percent) or that "they were only able to make a small contribution to their definition" (31 percent). This is to say that two-thirds of participants had little or no influence in the defi- nition of research topics. However, with respect to the methods applied in the projects, the Latin American leaders appear to have slightly more freedom to exert an influence: Almost half of them (52 percent) believed that they "took an active role in the design", while about another half of Latin American leaders (48 percent) maintained that they had little or no influence.

These data are corroborated, with slight variations, in regard to their role in the internal organization of the European projects in which they participate: Almost half of the leaders surveyed (48 percent) indicated that "the work was organized in an open manner, and that their group actively participated in the distribution of tasks", while slightly more than half experienced restrictions: 18 percent stated that "these were already defined previously and they had no involvement", while another 34 percent replied that they "had a very limited involvement". Regarding this important issue, we will see the contrast with the European point of view in the third section.

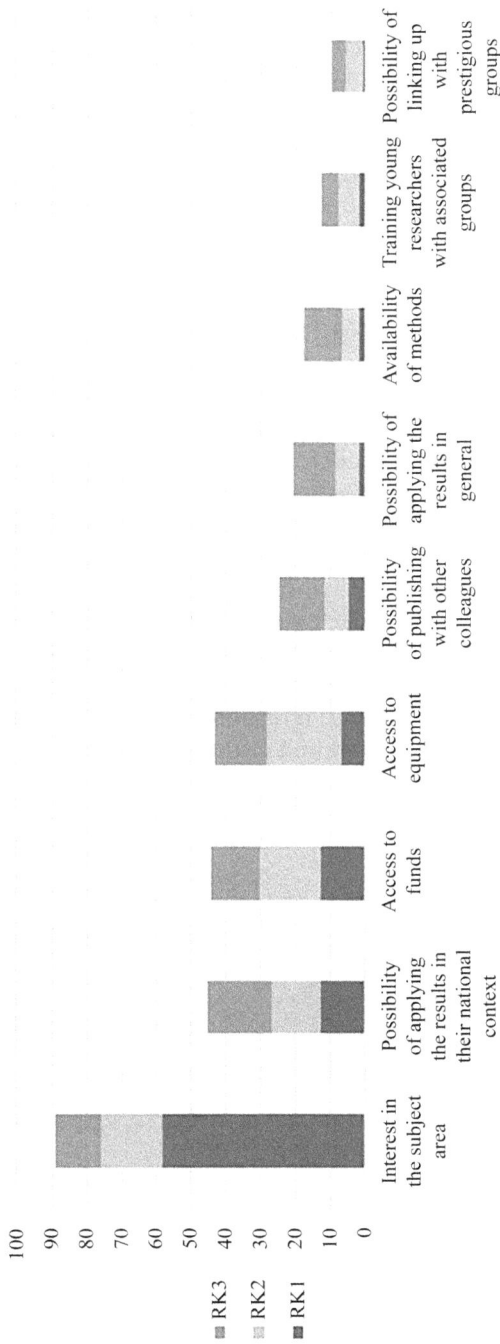

Figure 8.2 Motivation for Participating in European Projects[7]

Application/Orientation of the Knowledge and Results

Another aspect we looked at was the nature of the projects (basic research, applied research, or industrial development), the results expected and obtained, and their spheres of application. The data we obtained might be misleading (see Figure 8.3), since the majority of respondents replied that this knowledge was "directly geared towards industrial use" (18 percent) or else "could potentially have applications" (56 percent). The two answers combined make up almost three-quarters of the total, while only a quarter could be regarded as basic or fundamental research. However, we must make two qualifications here:

1. On the one hand, since the 1990s, most calls for projects, including those organized by both national policies as well as by international agencies, have in a sense "forced" or almost "obliged" researchers to mention the applications—whether real or potential—of their research (Joly, Rip, & Callon 2010). For this reason, when we analyze the distribution of projects between the formal categories fundamental research, applied research, or industrial development, the applied category is often heavily inflated.
2. As we have shown several years ago (Kreimer & Thomas 2005), in Latin America most of the projects labeled "applied" should be considered "applicable", since very few are actually applied in industrial uses, services, or new regulations. There are many reasons that can explain this fact, including most importantly the fictitious orientation in relation to the actors and institutions actually able to use such knowledge and the relative absence of institutions and firms able to industrialize the knowledge. The existence of high levels of "AKNA"

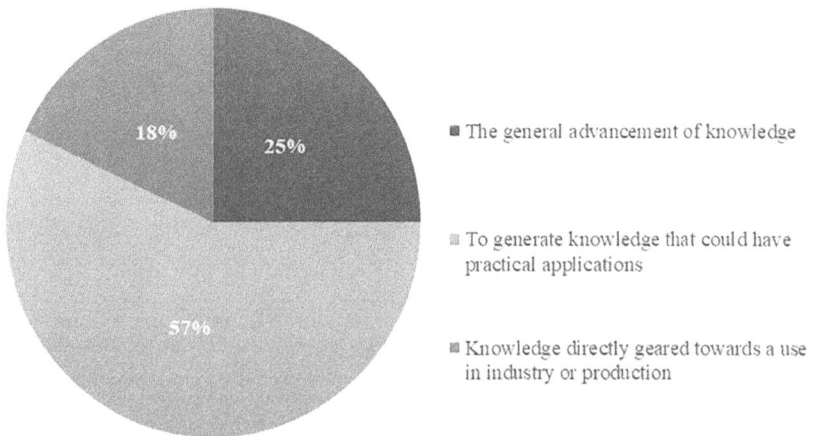

Figure 8.3 Main Orientation in the Application of Projects

(Applicable Knowledge Not Applied) can therefore function as a true mark of peripherality when analyzing the internationalization of science.

This is confirmed when we examine the results of cooperation projects (Figure 8.4): It is not surprising that approximately 45 percent of all respondents replied that they relate to publications (21 percent written by the group and 24 percent coauthored), while the "development of new industrial products" accounts for less than 10 percent. Another area of application involves generating "new national or international rules or regulations", which comes to 7 percent. In contrast, the acquisition of patents is barely represented. However, the other two statistically significant results relate to learning new technical skills (20 percent) and human resource training (18 percent).

Regarding the spheres of application (Figure 8.5) for the cooperation projects with Europe, the respondents gave one answer that was, at least in part, unexpected, since almost half (48 percent) replied that the sphere was Latin American countries and 27 percent Europe. The rest (23 percent) responded that the "sphere of the application was not defined", which may mean that these are basic science projects in disguise. However, this finding should be contrasted against the earlier-mentioned reflections on the research results, which qualifies to some extent the applied nature of the knowledge for Latin America: If only 17 percent reported having obtained applied results (including both "industrial developments" and the "generation of new regulations and standards"), then stating that they "apply the research results to Latin American needs" clearly seems to be more an expression of desire or an alignment with aims deemed to be politically correct than a verifiable practice.

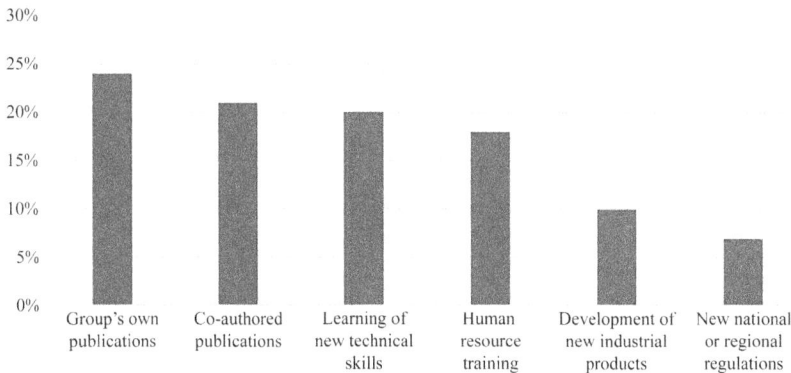

Figure 8.4 Results of the Cooperation Projects

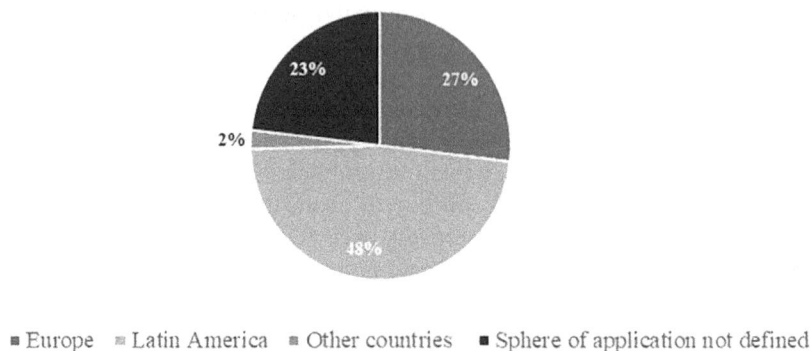

• Europe   ▪ Latin America   ▪ Other countries   ▪ Sphere of application not defined

Figure 8.5 Sphere of Application of the Project Results

Organization of the Work Within the Networks

The matter of the distribution of tasks within the networks is a crucial aspect for understanding the dynamics of this type of international cooperation and the functional differentiation that can be observed. What is striking in this regard is that the activity ranked first by the respondents (RK1) is "information/data collection" with almost a third of the total, and if "information/data processing" is added, the figure comes to 40 percent. Data processing tasks are also the option most often mentioned in ranking 2, suggesting that the two activities make up most of the tasks performed by the Latin American groups. We can add to this that "carrying out technical activities" also seems to be important, since it accounts for a third of the total number of responses in ranking 1, particularly innovative activities, with more than a quarter of responses, and routine activities, with just over 7 percent.

Thus, almost two-thirds of the activities performed by the Latin American groups in the networks are related to data collection and processing or to technical activities, while more creative activities, such as developing theories or new products or processes, appear to be relatively marginal practices, accounting for less than 10 percent of the responses. Of course, we cannot say whether *all* the members of the networks are engaged in the same type of tasks or whether this is a consequence of the organization of work in the networks, which are in fact highly heterogeneous in terms of both their disciplinary orientations and the kind of objectives pursued. At any rate, given the magnitude of the results, we can infer that the networks as a whole always involve a larger amount of more creative activities and that within this distribution the Latin American groups focus mainly on tasks related to data or to methods.

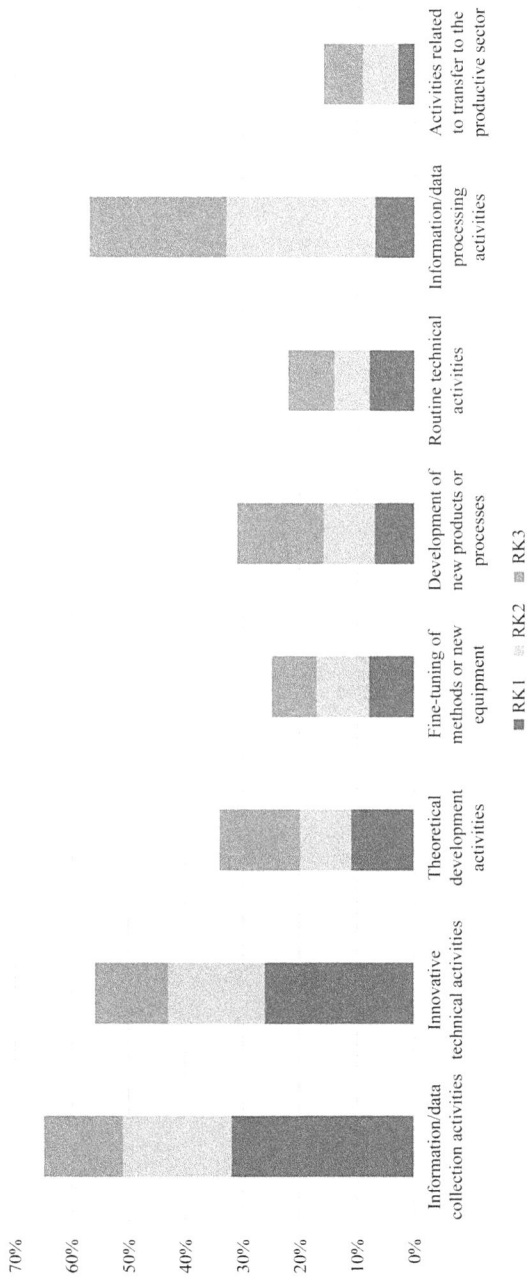

Figure 8.6 Types of Activities Developed by the Latin American Groups

3. European Views: Motivations, Perspectives, and Interests

Motivations for Enlisting Latin American Groups

The first thing we observe in European leaders' responses regarding motivations to include Latin American groups in consortia is that as a rule there exists a high valorization of the scientific quality of the Latin American scientists called upon. This was stated as the primary justification by all the European leaders interviewed. It should come as no surprise, for, as has been shown in previous chapters, elite groups from Latin America display a strong correlation in degrees of prestige and internationalization. Put another way, leading Latin American scientists build their reputations partly on local performance but above all by bringing their international relations to bear locally.

Even in the past, when the publication of scientific articles in peer-reviewed international journals had not yet been institutionalized as a standard measurement of scientific quality, a degree of local prestige was earned from recognition by international peers.[8] Today, participating in international and/or internationally funded projects is a source of prestige in all countries but is most apparent in those with intermediate levels of scientific development.

There are, however, substantive differences in terms of stimuli to enlist Latin American groups. We have identified four types of motivation/relationship. The references to the research consortia are coded to facilitate the reading, and the details of all the networks can be found in the appendix at the end of the chapter. Briefly, the areas considered are Health, Knowledge-Based Bio-Economy (KBBE), and Environment (ENV).

1. *The inclusion of Latin American countries as* a condition *for obtaining subsidies.*

This is the case with Health-5, KBBE-6, and ENV-6. In these cases, it is worth questioning the motivation behind the choice of *these groups over others* but not behind the inclusion per se of a Latin American group. The answer here seems to lie in two different sources: the first is technical and cognitive, like the additional capabilities provided by the Latin American groups invited or who have access to such cognitive resources as the availability of specific strains, access to patients, and so on; the second is sociological, referring to the linkages established between partners in the past, bonds of familiarity and trust, and even shared scientific paradigms, which are often mentioned by interviewees.

2. *Consortia working on* Latin American issues.

This is the case with most health projects, including tropical diseases, like Chagas disease: Health-1, 4, and 5. In this case we must pose an additional question: why Europe decides to fund research into issues that do not—or only marginally—affect its own context. Here again we find two different levels of response. The first relates to what research into local issues (diseases) can contribute to the understanding of more fundamental phenomena. In this respect, for example, the research into targets to attack the causative agents of these diseases may provide clues about fundamental physiological or biological mechanisms which can be extrapolated to other cognitive aspects beyond the cases in question. Moreover, testing brand-new molecules in association with pharmaceutical companies can produce knowledge applicable to universal diseases independently of exclusively local issues.

The second level of response relates to the outcome of processes of globalization and migration flows: Some traditionally "tropical" diseases have been spreading to other regions and creating fresh problems about which very little knowledge has been accumulated. This is true, for example, of Chagas disease in Spain, France, or the states of Texas and California in the United States. In these cases, on top of the prestige enjoyed by Latin American researchers, there is a specific expertise related to the object of study and exclusive access to essential research resources like patients or various strains of parasites and other organisms.

3. *Consortia investigating global problems with specific manifestations in a variety of contexts.*

There is a need here for various observation points located in different contexts. This is most often the case in environment-related projects and, in our study, in the ENV-1, 2, and 4 consortia. The phenomena investigated are due to more general phenomena like climate change and manifest in very specific geographical locations (the concentration of mercury in rivers or the consequences of melting of glaciers, for example). The incorporation of Latin American groups is due to two complementary modalities: the first is the need to collect data on specific observation points that produce a "global view" of the problem. Latin American groups must have access to those resources (rivers, glaciers, different species of fauna or flora, and so on) and the technical capacity to generate standardized data according to the consortium's protocols, since the use of the methodology here is key to the homogeneity of the data collected.

The second modality refers to the fact that the Latin American groups must be able to mobilize cognitive—and technical—resources, like knowledge of their specific contexts, and, above all, to be able to use equipment that allows them to carry out work consistent with that at the other observation sites.

Then the results are processed by program's coordinators (some projects have already undergone a degree of local processing) and the products are developed. These may take the form of intervention protocols, policy recommendations (to international agencies, the European Union or national governments), scientific papers, and, exceptionally, transfers to private companies, although in the case of environmental projects participation tends to be less than in other thematic areas.

4. *Latin America is an important—sometimes indispensable—context for observation and/or experimentation.*

This is true of the development of genetically modified organisms (GMOs) from the KBBE-1 consortium. For all that laboratory testing is authorized in Europe, it is virtually impossible to carry out field trials on any significant scale. In Latin America, on the other hand, there are very few restrictions on GMO testing provided certain basic technical norms are respected. It is then imperative for consortia performing activities prohibited or very difficult in Europe to acquire a field beyond their region in order to conduct such testing. Something similar occurs in KBBE-4 and Health-1 consortia, which involve clinical trials on patients which, although not banned in the European Union, are far more rigidly regulated than in Argentina and Brazil, where such testing has been carried out in both these consortia with less control and therefore speedier results.

Latin American Groups' Activities and Benefits

In our sample of research consortia, the predominant activities performed by Latin American groups in all thematic areas are *data collection and systematization* and *technical work*, be it routine or innovative. This refers us to a specific type of insertion in research planning consistent with the causes behind the recruitment of these groups: As we have seen, the predominant reasons in most projects are to do with observation sites of phenomena, either specific to Latin America (albeit with global interests) or local manifestations of phenomena far broader in their scope.

There are a further two facts of great importance: On the one hand, in the majority of cases Latin American groups were invited to participate in consortia *once the broad lines of the work plan had already been drawn up*, and their participation in the methodological design and distribution of tasks has therefore been extremely limited. There are only three exception cases of Health-1 and 4 and KBBE-6 where Latin American groups were actively engaged in the design. On the other hand, Latin American groups led the coordination of a working package, administrating some of the resources allocated to the project in just two of the consortiums analyzed. These are Health-4 and KBBE-6 (the only two projects where the groups participated in the overall research design), both cases involving international benchmarks FioCruz and EMBRAPA: highly prestigious groups from Brazil with very strong research traditions and high visibility.

Where company participation is concerned, European firms participate in all the consortia analyzed (with three exceptions: ENV-1; ENV-4, and ENV-5), while Latin American companies participate in two projects (one Chilean firm and one Argentine firm). Under European Union provisions companies involved in consortia must be small and medium-sized enterprises. Judging by consortia directors' statements, none of the companies participating in them has provided specific research funding; their contributions have instead been "in-kind": chemical compounds, data processing capacity, marketing and outreach, or specific equipment. We tried to investigate the nature of the industrial property and research results-exploitation agreements in all these projects, but it proved impossible because the provisions are confidential.

That said, we must highlight certain special characteristics displayed by different thematic and disciplinary fields:

Among health projects, two target Chagas disease, an endemic native to Latin America for which the most active research groups worldwide are found in Brazil, Argentina and to a lesser extent Colombia. The participation in the design of research and at the various different stages in these consortia is truly significant. One of them (Health-4) is one of the very few cases in which a Latin American group administrates a work package. Indeed, it is also one of the few to point out that the main benefit for the group is economic, since "the resources they administrate are very significant."

Similarly, all the health projects involve company participation, usually pharmaceutical laboratories, yet only one of these companies is Latin American. When we came to investigate who would be responsible for the industrialization of the results obtained from the consortium, we found out that a European industrial laboratory would be in charge of new drug production, while the Latin American laboratory would only see to distribution.

Moreover, in the area of health almost all the Latin American groups have links with patients and only in the case of Health-2 does the Latin American group (Chilean in this case) provide biologically interesting compounds and the capacity to carry out big data analysis.

We must not think that in the field of health data collection and systematization tasks are *purely technical* activities without any scientific content: Since it is about experimental developments, scientific capacities are crucial. These activities are, however, subordinated to a centralized information processing capacity always located in one of the European centers.

The situation in KBBE is fairly similar: It generally involves research processes in agriculture and agro-industry, where the availability of cognitive resources on the local environment and the Latin American groups' strong research traditions are decisive factors. Two of the projects are geared to conducting field trials in their own countries, these being banned in Europe (KBBE-1) in one case and regulations being slacker in our region (KBBE-4) in the other. Other groups also conducted clinical trials (KBBE-5) or field trials (KBBE-6), which are clearly their predominant activities.

Interestingly, the groups in KBBE that carry out more complex scientific activities are both from Chile: one is a veterinary center producing vaccines (KBBE-3), while the other is an expert in an international benchmark technique for bio-waste valorization (KBBE-2).

In the field of environmental research, whose tradition in Latin America is weaker than the two fields of health and agriculture discussed earlier, *all the groups identified perform information gathering and data production activities* and do so *according to already established protocols* across the board. The data generated by the Latin American groups target various different goals: systematizing the concentration of mercury in the water in various regions' (ENV-1), using global models with specific observation points (ENV-2 and ENV-4), producing conceptual models (ENV-3), or proposing policies and intervention tools (ENV-6). However, the participation of scientists from our region in these global objectives is either extremely low or nonexistent, as their practices often stop after the processing and systematization of the data produced.

In terms of research benefits for Latin American groups the findings are conclusive: The greatest benefit was seen in the increase in international publications. Added to this is the opportunity to interact with prestigious research groups at the international level. Contrary to expectations, according to their European coordinators, the securing of funding does not emerge as the main benefit for most Latin American groups. This may be for various reasons: First, as elite groups their basic funding needs may already be covered; second, there are *intangible* benefits like access to state-of-the-art methodologies, shared international databases, or other tacit knowledge; third, while the scope of the consortia's overall resources is relatively significant, the proportion received by Latin Americans (with the few exceptions where they are also administrators) is perceived as insignificant when weighed against other benefits.

4. Analysis, Comparison, and Conclusion

The first remark goes to confirm that the participation of Latin American scientific groups in European projects has increased significantly, encouraged in particular by the dynamics of the larger countries, which possess internationalized scientific elites, many of whose leaders have been trained or spent time in European institutions.[9] The rise of certain European countries as a destination for training thus seems to exert a strong influence on the type of links generated and their continuity over time, which then influences the decision to invite Latin American groups to participate in projects since, as we have seen (Figure 8.2), almost 60 percent of participations originate in this way, and another quarter arise thanks to "prior relationships" or informal contacts.

The main motivation appears to lie in the continuation of previous work undertaken by the Latin American groups, which leads us to observe that participation in international consortia is not, or is not perceived as, a driver for thematic innovation, but rather reinforces existing work. However, the general structure of the "thematic specialization patterns" of research studies (European Commission 2013) differ from one region to another, from which we can infer that there is a certain alignment between Latin American elites and European research patterns.

This also downplays the role of the Latin American S&T policies, insofar as their countries' scientific elites gear themselves more towards topics on which they can collaborate with international peers, than towards more "purely local" issues. In this regard, it may seem paradoxical that part of the growth in Latin American participation in European projects could also be explained by the impetus provided by the international scientific cooperation policies of several Latin American countries: establishment of contact points with Europe, availability of more counterpart funds (which is common to all the countries analyzed).[10] However, national policies actively served to encourage (and finance) the participation of their researchers in European projects but had no role in the thematic orientation of the networks and projects, which continued to be guided by a markedly laissez-faire approach.

Through our empirical analysis, we have been able to observe that, whereas scientific relations are becoming more complex and Latin American research groups' participation in international consortia is on the rise, the basic structure of these relations is still organized around "subordinate integration" modalities. The activities most frequently undertaken by Latin American researchers in the research consortia's division of labor are data production, organization and systematization. The spread of technologies for sharing both materials and research results does not seem to have altered the structure of previous relationships. Meanwhile, with few exceptions, European groups tend to concentrate on research design—in both theoretical and particularly methodological terms—and the ability to centralize any data generated and to produce conceptual interpretations.

Given that almost all Latin American research groups (with the earlier noted exceptions) are invited to participate in consortia once the research design has already been established, their capacity to steer the results of their work towards the needs of or potential use in Latin America is extremely low.

From the Latin American point of view, a crucial motivation of participation is related to the extension of links with prestigious European groups and the possibilities of publication. This coincides with the perspective of researchers 40 percent of whom state that the results of collaboration are published papers. However, in line with the motivations expressed, this increase would not appear to reflect a greater thematic variety, but merely a rise in the number of articles on topics existing prior to the implementation of the projects.

From a European point of view, the nature of the collaborations appears to rest on the availability of Latin American researchers, with whom they already share a common (scientific) language and who can make significant contributions, particularly regarding technical issues or, above all, information collection and processing. At the same time, the Europeans seem to reserve tasks related to theoretical production for their own groups, since these occupy a marginal place (just 10 percent) for Latin Americans.

This leads us to wonder about the use of the knowledge generated in the networks. As we have shown, the researchers themselves state that around a quarter of research is classified as "science for the sake of science", while 56 percent indicate that, potentially, it "could have an application". Given the complete absence of declared patents, our analysis of publications confirms two trends: one, towards the "fundamentalization" of research as demonstrated by the evolution of clusters observed in an emerging discipline such as the nanosciences (Bozeman et al. 2007; Klein 2011); the other towards increasingly global issues, such as the evolution of topics in medicine. In this context, there seems to be very little room—in spite of what the regional leaders themselves state—for participation in European programs to produce applied research to address local issues.

Furthermore, as we pointed out in the last section, European companies, usually medium-sized enterprises, participate in almost all consortia. Although for reasons of confidentiality we were unable to access industrial property and results-exploitation agreements, it is reasonably safe to infer that, should commercially viable results be obtained, they will be used first by companies participating in these consortia and swiftly thereafter by other European companies. The industrialization of knowledge—if it takes place—will be deployed primarily on the European continent.

The findings reinforce some previous studies that speak of a division of scientific labor between hegemonic and peripheral groups in which the latter tend to carry out technical activities or those related to data collection and processing (often in relation to local resources, such as certain tropical parasites in medicine, soil samples in geology, or the activity of Antarctic fish in marine biology), while the former usually have a far more active role in data monitoring, synthesis activities, and conceptual output (Kreimer 2010c).

The international division of scientific work may be explained—at least partially—by various causes. First, the S&T policies undertaken by developed countries, characterized by a significant increase and concentration of resources, aimed to generate "large blocks of knowledge" like the European Research Area (ERA).[11] Second, the increase of communications by electronic means (ICTs) reinforced the intensity of scientific collaboration, including the creation of "virtual laboratories". The growth of scientific human resources in Latin American countries and their policies that encourage international cooperation are also factors to be considered.

All these causes could explain the increase or intensity of the cooperation, but not the specific shape that we systematically found. The current study leads us to think that is the structure of international cooperation who is modeling this specific shape.

Looking at the variables that could explain the increase of collaboration and explaining how consortia are put together (Q1) we can confirm our hypothesis: There is indeed a reciprocal functionality. But while Latin American groups look to publish while EU groups has broader interests.

About the consequences of this specific type of scientific collaboration (Q2), we show that while Latin American scientists believe that there is no change in scientific agendas, we can observe a slightly process that undermine local issues in favor to international ones. Of course, the intensity of the production of knowledge is increasing, but this has no effect on agendas (or the effect is not controlled from Latin American countries).

This type of collaboration creates the "fiction of autonomy" in relation to the local contexts of work and a sort of new global "republic of science". This situation seems to incorporate some kind of "democratization" in the relationships that govern the production of knowledge, within the frame of "universalized" links, close to Wagner's thesis.

However, it causes different effects for developed and developing countries. In fact, as we saw, the groups located in nonhegemonic contexts basically conduct research practices with relatively low theoretical content, with a predominance of data collecting and processing tasks of, as well as mostly technical work. Thus, part of the research work is transferred or "outsourced" from developed to developing countries, while the cognitive control of research is managed from the core groups within each subject area or discipline. This confirm our second hypothesis (H2)

It must be added to the former statement that companies usually participate in and cofinance the development of these projects (as it is explicitly stated in several call for projects), who industrialize useful research results. But these companies are usually located in developed countries, so this fact implies that it is practically impossible to orient research results to tackle local issues in developing countries. This reinforces the hypothesis of reciprocal functionality that we enunciated at the start, insofar as these interactions give rise to a more sought-after product: the publication of co-authored papers.

We may conclude that it is evident that Latin American elite research—in larger countries—is actively integrated into international networks and global science, accompanying the globalization process that arose during last decades, and the growth of the intensity of communications, relationships and sociocognitive links. Nonetheless, some kind of "neo-colonialist" trend may be perceived when observing the changing orientation of research agendas, the unequal distribution of task within each network, and the uneven effective use of knowledge produced as a result of collaborative projects.

At this point we might hypothesize the existence of a "functional inter-dependence" between European policies to stimulate the participation of extra-European scientists and Latin American elites: Whereas the European Union effectively expands the number of researchers tackling specific questions of interest to Europe, Latin American elites find a means to formalize their international linkages and to increase their publications and the general international visibility of their work.

Let us look again at the two paradigms that form part of the European Union's explicit discourses: the narrow and broad paradigms. Based on our research, the implementation of the narrow paradigm—namely, the aim "to improve the quality, scope and critical mass of research"—has involved participation by a growing number of Latin American researchers who provide specialized research capacities in specific thematic fields, which are then added to any specifically European human resources. Unlike in the past, this takes place without any permanent scientific or long-term migrations; instead, Latin American groups work from their own countries and provide the infrastructure already existing there for that purpose.

Moreover, an effect of cultural proximity and extended communities is produced among peers from the two regions, swelling the amount of knowledge available for potential use by Europe.

It is worth considering the "broad paradigm"—namely, the policy level and the aim to improve competitiveness, tackle global social challenges, and support the least developed countries in science and technology capacity building. Briefly, given the features of the significant development of scientific groups in the countries in question, the strengthening of these countries' capacities is undoubtedly the result of the cooperation analyzed but is not a decisive factor (as it would be for less developed countries). On the other hand, the nature of the projects considered and their dynamics clearly helps to strengthen the European Union's global view due both to "observation points" of various global phenomena *in* Latin America (like climate change) and the opportunity for testing in less regulated fields, and even to the opportunity for generating local knowledge to complete a broader picture or contribute to the production of regulations in Europe.

Regarding both points of view, Latin American and European, we can suggest a typology (certainly not exhaustive) of the various modalities of Latin American groups' insertion in international consortia based on our empirical work. For this purpose, we identify four types of participation and take some of the consortia analyzed as concrete examples to illustrate them.

- Type 1: Participation based on local resources in Latin America essential to the research.
- Type 2: Participation via "outsourcing" for specific contributions to some global research or to produce regulations.
- Type 3: Participation based on the international reputation of the group in Latin America or specific knowledge needed for the research.
- Type 4: Participation based on the opportunity to conduct (clinical, agricultural) trials that are less regulated in Latin America than in Europe.

In order to briefly illustrate this typology, we can see some details of four consortia that show the nature of the scientific relationships:

Type 1: BERENICE (Benznidazole and Triazole Research Group for nanomedicine and Innovation on Chagas Disease)

A consortium coordinated by a Catalan group to test the toxic profile of benznidazole, the only drug in existence to treat Chagas disease.

The Brazilian group (FioCruz and Ouro Preto) devote their energies to designing and conducting clinical trials. According to the coordinator "they are a key piece: the trial is based exclusively on the data produced by the Brazilian groups (with Chagas patients)". They are also responsible for in vivo models. The Argentine group (ANLIS) is responsible for testing new alternative molecules to benznidazole, and their contribution is also "indispensable thanks to their experience with *T. cruzi*".

In spite of their apparent importance, the Latin American groups played no part in the project definition (objectives, methods or distribution of tasks).

Type 2: VIROCLIME (Impacts of Climate Change on the Transport, Fate and Risk Management of Viral Pathogens in Water)

A consortium coordinated by an English group to analyze the contamination of river water by different viruses as a result of climate change.

The project studies various European rivers and contrasts them with a tropical region (studies in the Amazon basin). The Brazilian group UFRJ was called because "it is highly respected internationally", but mainly because "it was necessary (and timely) to have a tropical location, and they have access to the Amazon". Its job was "to take samples according to the protocols, make the measurements—which is a delicate task—and produce the relevant reports".

The coordinating group had already had two requests from the European Commission (and two European projects) to prepare a policy report on the epidemiology of the rivers and the measurements of viral concentrations.

Type 3: EPIMIRNA (MicroRNAs in the Pathogenesis, Treatment and Prevention of Epilepsy)

A consortium coordinated by an Irish group to find a treatment using microRNA (small cellular molecules) for cases in which traditional medicines are ineffective in treating epilepsy.

The Brazilian group (Campinas) was called because—according to the European coordinator, it is "very well-known internationally, with important publications in the field of epilepsy. Above all, they are very strong in genetic aspects and what we wanted in this project was to look at the genetics of a particular type of gene in epilepsy. This Brazilian group really has incredible competences in the field of the genetics of epilepsy".

On the other hand, it was very useful for the consortium to carry out testing on "other populations—genetically speaking—outside Europe [. . .] which is why we imagined running trials in the USA and Brazil" (the group is linked to a clinic in São Paulo, where patients are available). The genetic work outside Europe, however, was coordinated by an American researcher from Columbia University.

Type 4: AMIGA (Assessing and Monitoring the Impacts of Genetically Modified Plants on Agro-ecosystems)

A consortium coordinated by an Italian group to assess the environmental and economic impact of GMO crops. One of crops assessed in this project is the genetically engineered potato, a topic on which the Argentine group (INTA) has been working for several decades. The Argentine group's particular task is to work toward the validation of the control methodology in areas where GMO crops are grown on a large scale. A first validation of the methods is foreseen in the United Kingdom, but it needs to be tested in large-scale production, something which is permitted in Latin America but not in Europe.

In closing, I hope the (preliminary) typology proposed will provide a springboard for further analysis of Latin American groups' participation in international consortia and, more generally, of certain modalities upon which the relations between hegemonic centers of knowledge production and semiperipheral contexts are structured. I also think the typology must be enriched with new empirical researches that, alongside cooperation with Europe, take account of the linkages funded by the various agencies of the United States and also by less formalized cooperation modalities in order to detect whether they follow a similar pattern to those we see here or whether any distinct modalities can be identified.

Appendix 1
Methods

1. Latin American leaders' opinions:

In February 2015 we sent an e-mail questionnaire to 910 Latin American Countries (LAC) project coordinators. We received 210 complete answers (23 percent). Statistical analysis regarding the distribution over countries and Framework Programs shows that our only underrepresented data was Brazil in FP7, with a slightly 10 percent less than expected. The questionnaire consisted of 21 questions: (a) a set describing the Latin American group involved in the project; (b) questions dealing with the specific way in which the collaborative relationship was established, affinity to project topics, involvement in setting research agendas, methods and orientation; (c) a set of questions that included the assessment of the sphere of application and motivations, the organization of work within the network, and the specific type of activities carried out in Latin America; and (d) questions on the LAC group's contributions and the results obtained from the collaboration. Since the responses from those who participated in FP6 or FP7 did not show significant differences we examine them all together here.

2. European research coordinators opinions:

The empirical work was oriented to carry out qualitative research about the motivations of European leaders to enlist Latin American groups in European projects and the type of practices they undertake within large consortia. Therefore, it has no statistical validity, nor are its results generalizable to all scientific cooperation activities. Instead, we chose to observe paradigmatic cases that provide us with more detailed information about the object of study. The selection of cases that are studied in depth are not intended to be representative, but rather open the question to various possibilities and relations of the actors, their traditions, disciplines, working subjects, possible industrial applications, etc.

It was based on an analysis of subprojects of the cooperation sub-program of FP7. First, we selected from the CORDIS database projects with participation by at least one research group from Argentina, Brazil,

Mexico, Chile, or Colombia. Second, we organized this sample thematically, selecting areas where Latin America has both a greater research tradition (Health and KBBE) and a different research tradition, which unlike previous ones involves on-the-ground as opposed to laboratory research, chiefly aimed at studying global systems (ENV). Next, between five and six projects meeting the following requirements were selected from each of the three areas: being financed by the "collaborative projects" scheme[12] and addressing different topics within the area in order to represent projects with a diversity of orientations.

The characteristics of the selected projects have been studied through two mechanisms. First, the website of each project has been explored (seventeen in total), which includes the coordinators, partners from different countries and institutions, the general description of the project (including the objectives of each working package), the activities carried out, and the results obtained. In a second time, we conducted a semistructured interview with the coordinators of the selected projects, to explore the following issues:

a. The origin of the project (coordinator or other group) and participation of Latin American groups during the formulation process.
b. The motivation to specifically convene these Latin American groups: advantages in terms of evaluation, the existence of highly trained human resources, access to resources in the Latin American region (specimens, populations), knowledge of technical or specific areas, counterpart funding contributions, and so on.
c. The distribution of work packages.
d. The type of activities carried out by the Latin American groups: routine activities, innovative techniques, the development of new techniques and equipment, information or data collection, and theoretical development, development of new processes or products, productive sector transfer.
e. The results obtained: joint publications, patent registration, new regulations, standards or norms.
f. The roles and characteristics of companies comprising the consortium.
g. Negotiations around the intellectual property of the possible outcomes.
h. Benefits for Latin American groups: human resources training, the opening of new lines of research, access to funding, contact with international scientific leaders and integration in global networks, the meeting of social needs, and the creation of new companies.

The basic features of the consortiums analyzed are set out in Appendix II, where we have summarized information on the objectives, the participant groups, European leaders' motivations, and the activities carried out by the Latin American groups in the three areas chosen: Health, Knowledge-Based Bio-Economy (KBBE), and Environment (ENV).

Appendix II
List of Investigated Consortia

Table 8.1 Sample of selected consortia involving at least one research group from Argentina, Brazil, Chile, Colombia, or Mexico

Project acronym	Thematic area	Main objective	Coordinating country	Participant LA country	WP country	Company participation	Main benefit to GL	Primary motivation to include GL	GL task
BERENICE (HEALTH-1)	Health	To improve existing drugs for Chagas Disease with nanomedicine	Spain	Argentina (2 groups: Fatala, Chaben, ELEA company) Brazil (FioCruz)	NO	YES	Access to international networks	Previous experience of both groups Argentine groups have *patients* Brazilian groups have *in-vitro testing*	Clinical trials and in-vitro testing
DIVINOCELL (HEALTH-2)	Health	To develop new components for treating infections caused by gram-negative pathogens	Spain	Chile (University of Chile)	NO	YES	Significant resources	Properties of certain target compounds	Composites testing. Highly specialized technical data analysis
EPIMIRNA (HEALTH-3)	Health	To gain fresh understanding of and develop treatment for epilepsy via microRNA	Ireland	Brazil (UNICAMP)	NO	YES	Development of methods totally new to LA Publications Incorporation in international consortia	High prestige Has clinic and patients	Clinical trials with Pre-designed Protocols They collect, produce and analyse data
PARADDISE (HEALTH-4)	Health	To test the findings of the targets and molecules for three parasitic diseases: Chagas' Disease, malaria and leishmaniasis	France	Brazil (4 groups: FioCruz, USP, UFRJ and UF de Viçosa)	YES (FioCruz)	YES	Financing is extremely important for FioCruz and the USP	Need to work on endemic areas Leading-edge expertise of Brazilian groups Availability of biological material (parasites)	Big Data production and processing
SETTREND (HEALTH-5)	Health	To describe new therapeutic targets for schistosomiasis	France	Brazil (4 groups: FioCruz, USP, UFRJ and UF de Viçosa)	NO	YES	Publications and some financing	Need for groups from endemic areas International expertise	Rio de Janeiro: Bioinformatics validation of targets São Paulo: Transcriptomics

(Continued)

Table 8.1 (Continued)

Project acronym	Thematic area	Main objective	Coordinating country	Participant LA country	WP country	Company participation	Main benefit to GL	Primary motivation to include GL	GL task
AMIGA (KBBE-1)	KBBE	To use model cultures to design biosecurity protocols (assessment methodologies) for GMOs	Italy	Argentina (INTA)	NO	YES	Opening of new lines Access to international networks	The group's skills/ expertise in GMOs The possibility of conducting field tests (banned in Europe)	*Monitoring; evaluating plant and insect resistance; evaluating whether resistance is exclusive to a region. Sophisticated technical work*
GRAIL (KBBE-2)	KBBE	Biowaste recovery in biorefineries	Spain	Chile (Pontifical Catholic University of Valparaiso)	NO	YES	Significant resources Access to international networks	High prestige group Expertise in industrial biotechnology Use of an international benchmark technique	Innovative technical activities and development of new processes and products
TARGETFISH (KBBE-3)	KBBE	To develop new antigens and improve existing ones to design vaccines for diseases of farmed (bred) fish	Netherlands	Chile (Centro Vet, company)	NO	YES (50 % of total institutions)	Interacting with global universities and companies Product placement in European market	Expertise in vaccine production	Product development
PLANTLIBRA (KBBE-4)	KBBE	To carry out risk assessment on plant-based food supplements Chemical and clinical toxicity analysis	Italy	Argentina (UBA) Brazil (USP)	NO	YES	Argentine group: resources, materials Brazilian group: international networks Both: publications	Potential to conduct less regulated clinical trials	Data collection and processing

LOWINPUT-BREED (KBBE-5)	KBBE	To improve animal health in terms of the quality of European organic and 'low impact' products in milk, eggs and meat	United Kingdom / Switzerland (scientific coordinator)	Brazil (UF de Viçosa)	NO	YES	Training human resources and incorporating new techniques	Different varieties of pigs, as they worked with many countries, and this added diversity to them.	Innovative technical works
SWEETFUEL (KBBE-6)	KBBE	To optimize sorghum yields in semi-arid, temperate and sub-tropical zones, for genetic improvement and agricultural practices	France (CIRAD)	Brazil (EMBRAPA) Mexico (Nuevo León)	YES (EMBRAPA)	YES	Resources, international networks	Brazil: necessary for the call and a benchmark in biofuel Mexico: potential pilot plant	Brazil: Field tests Mexico: Pilot plant
GMOS (ENV-1)	Environment	To produce data on mercury concentrations in the environment and in fluids across the world	Italy	Argentina (INIBIOMA) Brazil (USP-LBA)	NO	NO	Equipment, resources, HR training, publications	Observation points in South America	Field data collection and analysis from GMOS-established protocols
ACQWA (ENV-2)	Environment	To use models to quantify the influence of climate change (glaciers) on the determinants of mountain river flows (impact on agriculture, tourism and hydropower)	Switzerland	Argentina (ITDT, economy) Chile (2 groups: La Serena, Valdivia)	NO	YES (hydroelectrics)	La Serena and Buenos Aires: resources and international networks Valdivia: strengthening its networks	Comparing the Andes with other regions Valdivia: high prestige and great international expertise	Sample collection, data production

(Continued)

Table 8.1 (Continued)

Project acronym	Thematic area	Main objective	Coordinating country	Participant LA country	WP country	Company participation	Main benefit to GL	Primary motivation to include GL	GL task
SPECS (ENV-3)	Environment	Conceptual structure to join up climate predictions and information services with users (from policy makers to industry)	Spain	Brazil (INPE)	YES	YES	Human resources training Strengthening international networks Resources	High prestige group with great expertise	Innovative technical activities Data processing and analysis
VIROCLIME (ENV-4)	Environment	To use hydrological models to determine the effects of climate change on viral variation in water and to establish risk of disease	United Kingdom	Brazil (FioCruz)	NO	NO	Publications Strengthening international networks	High prestige group Access to natural resources (Río Negro)	Innovative technical activities Data processing and analysis
THESEUS (ENV-5)	Environment	Comparative studies on "climate proof technology" in coastal areas: risk assessment and mitigation policy proposals	Italy	Mexico (UNAM)	NO	NO	Publications Access to international networks HR training	Access to local natural resources (Cancún)	Innovative technical activities
FORCE (ENV-6)	Environment	To study the causes of the change in the coral reef ecosystem To propose management tools: policies, governance frameworks, human behavior control, regulations	United Kingdom	Mexico (2 groups: UNAM and Colegio de la Frontera Sur)	NO	YES	Resources Publications	EU call requirement Access to local population and natural resources	Data collection Innovative technical activities

Notes

1. To facilitate the reading, at the end of the chapter the methods used to gather the respective opinions of Latin American and European scientists are described.
2. For a discussion of the inadequacy of assimilating "North" to developed countries and "South" to developing ones, see Chapters 2 and 7.
3. This finding is consistent with Adams's assertion (2012) that small countries like Malta find it impossible to identify what "own science" is.
4. This figure only includes Latin American groups' participation in research projects, not researcher mobility projects.
5. In fact, in terms of human resources the number of researchers in relation to the economically active population was lower in Europe than in Japan or the United States: in EU-25 in 2003 the ratio was 5.4 per 1,000 as against almost double that in the United States and Japan (9 and 10.1 per 1,000 respectively). Of course, there are significant differences within Europe (Muldur et al., 2006).
6. To give an idea of one of these structural asymmetries, it should be remembered that in countries like Argentina and Brazil the participation of private for-profit institutions, added to that of small and medium-sized enterprises, accounted for about 14 percent of total participation in PM7.
7. In this question the respondents were asked to select a maximum of five options with the most frequent activity in first place (RK1) and the others in descending order. We have shown the results of just the first three options, which have been weighted according to their relative importance.
8. For case histories of this process of external scientific capital building, see Romero (2004), Cukierman (2007), Buch (2006), Kreimer (2016), among others.
9. Gérard and Cornu (2013) show, for example, that countries such as Spain, France, and the United Kingdom (and to a lesser extent Germany) are increasingly important destinations in the training of researchers: While in the 1970s nearly half went to the United States, in the 2000s almost two-thirds of Mexican researchers undertook training in one of these four European countries. Similar trends are observed in Argentina, Brazil, and Chile, albeit with significant disciplinary variations.
10. This is evident in the case of Argentina, Brazil, and Mexico, the latter of which had lesser relative participation up to FP6 and provided the greatest impetus through its policies, in order to both balance the ever-important influence of the United States, as well as to catch up with the other active countries in the region.
11. One of the objectives of the ERA is to "develop close ties with partners around the world so that *Europe benefits from the worldwide progress of knowledge,* contributes to global development and take an important role in international efforts aimed at resolving issues of world importance" (European Commission, 2007, my emphasis)
12. We focus on schemes that back strictly research projects and disregard financing for establishing and/or strengthening international networks, where Latin American participation is less relevant.

9 International Scientific Collaborations at the End of the World

Local Resources and Global Research in Tierra del Fuego

Introduction

In the last two chapters, I looked at various aspects of "center–periphery" relations, noting certain milestones in their historical development and their consequences in international scientific cooperation processes (especially in Latin American researchers' participation in European projects). Going deeper in that direction, in this chapter, I want to present a specific case set in the "the world's southernmost center"[1] and inquire into the structure, content, and direction of scientific activities carried out in cooperation in a far more globalized world than Jules Verne's, a little over a century on.

The Austral Center of Scientific Research (CADIC) of the National Council for Scientific and Technological Research (CONICET) is a geographically remote center of Argentine scientific research located in Ushuaia on the Beagle Channel, in the Tierra del Fuego archipelago off the southernmost tip of the South American mainland. Research activities of this center belong to several disciplines: marine biology, geology, terrestrial ecology, and archeology. Ushuaia is only 1,000 kilometers away from Antarctica, while its distance to the most dynamic spaces of knowledge production in Argentina is more than triple that figure.

Perhaps contrary to expectations over such a remote research center, the CADIC has been intensely involved in international scientific collaborations. This rather surprising finding raises several interesting questions: Why has a geographically distant center attracted the interest of foreign investigators, to the point of exhibiting one of the highest rates of international collaborations per researcher compared to other scientific institutions in the same country? What kind of resources does the region offer in the center's scientific fields that might explain this phenomenon? What patterns of exchange and what tensions emerge? How does the changing international scenario affect the current nature of scientific exchanges among research fields?

To suggest some answers to these questions, I analyzed more than fifty cases of international collaborations conducted by CADIC researchers in the fields of marine biology, geology, and terrestrial ecology using a questionnaire and interviews with the participating scientists and PhD students.

As I have shown in previous chapters, Latin American countries have been significant contributors to the growth of some global research networks (Leydesdorff & Wagner 2008). Yet little is known about the patterns of collaboration that arise or the motivations that lead researchers in scientifically advanced countries to associate with those pertaining to developing or peripheral countries or about the effects of these changes on the least scientifically developed countries. While many observers in Latin America tend to emphasize the synergic effects of international collaborations and their benefits to the region, others have underlined certain asymmetric effects of this growing participation of "peripheral", or "nonhegemonic", countries.[2]

Of course, the primary motivations for international collaborations are supposed to vary with the features of the scientific fields involved and the topics approached. Those performed in the CADIC—included in my research—have a common matrix: They focus on certain natural features with an uneven geographical distribution and, therefore, access to those *resources* is required.

I open with a discussion of a few features of these scientific fields that have an impact on patterns of international collaboration. There follows a brief outline of collaborations in the CADIC, and, lastly, examples will be offered to shed light on the nature of the interests that attract foreign scientists to associate with CADIC researchers, and on the role played by local resources and knowledge. In conclusion, I make some remarks on internationalization in the fields under study.

1. Scientific Fields Dependent on "the Geography of the Research Object": Some Characteristics Relevant to Patterns of International Collaboration

We focused our research on marine biology, geology, and ecology. These fields—and some other disciplines not here considered—share one feature that has a strong influence on patterns of international scientific collaboration: they have a sort of "geographic dependence", that is, their research problems require access to certain resources not universally available due to uneven distribution across the world. "Things" like glacial or volcanic landscapes, or certain species of flora and fauna, certainly cannot be found in all regions. Differentiating motivations for international collaboration in science, Wagner (2005) refers to these fields as being "resource-driven", and Jappe-Heinze (2007) talks about "the geography of the research object" and how it relates to patterns of collaboration. However, the term "resource" is potentially misleading, as is the term "object", used to refer to regional peculiarities relevant to scientists in a given field. For want of a more precise word, both expressions are used here on the understanding that, even if specific regional characteristics are, in some way, provided by nature, they do not become an "object" or "resource", except in the context of sociocognitive organization (see Chapters 3 and 4)

Since this geographical dependence on the patterns of international scientific collaboration have not received much academic attention until recently (Jappe-Heinze 2007), I summarize here some of the consequences, based mainly on the case study. While some of our comments may only apply to our case, others undoubtedly go beyond this context. These features are later used as explanatory factors, among others, in a brief description of the CADIC's patterns of international collaboration and are accompanied by some examples.

a) *Geographical Advantage: When Local Peculiarities Have Something to Tell Us About Global and/or General Issues*

The potential of a regional resource to attract scientists from other countries depends on the relevance of the research questions it may help to answer from the viewpoint of foreign scientists and groups. So, not all regional peculiarities will become a "geographical advantage" for local scientists. In our case study, for instance, geology and marine biology seem to frequently enjoy such advantages in Tierra del Fuego, but this situation is highly unusual or even nonexistent for forestry research. Local resources attract the attention of prestigious researchers (those who have accumulated scientific capital) insofar as they are seen as "relevant problems" whose importance transcends the local scenario.

According to our research, local conditions may have this kind of relevance in two types of situations. The first type is somewhat obvious: If certain "phenomena" only take place or can be seen in few geographical areas, then accessing these areas will be mandatory for research if one wishes to describe those phenomena or postulate general explanations on an empirical basis. This may be the case with certain unusual geological formations in terms of the mechanisms that gave rise to them, and it is also the case with certain biological species that can only be found in given locations. In such cases, conclusions do not need to be limited to the species under study. They could shed light, for instance, on more general physiological processes, or observed biological adaptations could, at least in principle, suggest medical applications. Furthermore, the "phenomenon" under study does not always have to be infrequent or "confined" to certain areas. It may be difficult to obtain samples of the desired quality, and therefore some regions seem to provide this more than others.

The second type is more complex, involving "local manifestations of global phenomena". Indeed, the fields we are focus on here are not only concerned with establishing generalizations, as they deal with particular configurations of the Earth in different times and spaces, and therefore with particular processes of change. Although certain subfields of geology perhaps furnish the most obvious examples, global climate change has, in the present, become the object of study in many fields, providing an outstanding example of this kind of search. The study of processes involves consideration of different systemically related factors operating frequently on large scales (which may be, but are not necessarily, "global", as in current climate change), whereas regional manifestations of a process may vary. Nevertheless, some can be deemed more relevant than others in shedding light on the entire process. This is the case, in our own day, with the Antarctic—and sometimes the sub-Antarctic—region for the study of global climate change. Scientific interest in a region may, therefore, depend on the importance of the contribution it is expected to make to the understanding of a process that takes place on larger (frequently global) scales.

Though many studies combine these two kinds of concerns in different ways, the distinction is useful to clarify the sort of cognitive interests that attract foreign scientists, and when presenting some of our cases, we will have occasion to illustrate how regional resources can be used with these different aims.

b) Local Researchers as "Obligatory Passage Points"? Access to Local Resources Under Dispute

For local scientists, having privileged access to a *resource* (in the earlier meaning) implies some sort of monopoly on certain "research objects" that can, in principle, facilitate links with researchers from scientifically

Table 9.1 Patagonian expeditions

From the early nineteenth century, Southern Patagonia was the object of several scientific expeditions, not to mention literary imagery, like the book by Verne. Scientific expeditions were numerous, notably Alexander von Humboldt's between 1799 and 1804, a true milestone in our knowledge of the region.

Then, of course, there were those organized by Phillip Parker King between 1826 and 1830, and continued by Robert Fitz Roy in 1832 and 1836, particularly relevant as it carried Charles Darwin, who gathered information crucial to his work.

Also worth mentioning is the journey of adventurer George Chaworth Musters between 1869 and 1870, based on which he published a charming book entitled *At Home with the Patagonians* (1871).

Some years later, three scientific expeditions to Patagonia were organized by Princeton University, between 1896 and 1899. The primary focus was to collect vertebrate and invertebrate fossils. The journeys were inspired by a series of well-publicized paleontological discoveries by Drs. Florentino and Carlos Ameghino, beginning in 1887. Expedition members collected a wide range of local flora and fauna, including birds, plants, mammals, amphibians, freshwater fish, and mollusks.

advanced countries. However, this statement needs to be nuanced and elucidated further.

The monopoly enjoyed by local scientists is relative. Researchers from around the world may find the possibilities offered by a given region attractive, and this will not always imply that local scientists act as "obligatory passage points" (Callon 1986). Access for scientists from abroad to regional resources depends, among other factors, on the clearance required to conduct research in a certain territory. According to Wagner et al. (2001, p. 49), "in some countries, either explicit or *de facto* government policy may require cooperation with a national to gain access to these types of local resources". Nevertheless, this is not always the case. In Argentina, foreign geologists need no special clearance to conduct fieldwork, and geological travelers are, indeed, a common sight in Patagonia and Tierra del Fuego, as reflected in numerous papers based on data collected in the region by foreign scientists.

In marine biology, even if the consent of coastal states is required for research ships, the local monopoly is weakened by facilities and technologies available today not only for sample conservation and transportation but also for transportation of live specimens (such as deep sea or Antarctic and sub-Antarctic species) that require special, "extreme" conditions to survive. Therefore, once collected, these specimens can be studied outside their habitats ("purified")[3] for many research purposes. Over the last two decades, Antarctic research vessels have extended their search to the southern tip of South America. Nowadays, with the aid of such vessels, it is possible to perform large-scale sampling of species in Antarctic and sub-Antarctic regions. Experimental research can sometimes be done on board, but, most importantly, collected samples and live specimens can reach northern laboratories, something that was not always possible in the past, as keeping alive specimens that require special conditions to survive is no easy task.[4]

c) *The Role of Local Knowledge*

Unlike other scientific fields focusing on aspects of the world that are not expected to differ according to the location where research is performed, the regional specificity of the phenomena under study in fields with a "geographical dependence" is crucial. And, for scientific research, this entails the requirement of producing *site-specific* knowledge, be it about local flora or fauna, the region's geological history, or something else. Obvious as it may seem, this fact has consequences in terms of patterns of scientific collaboration because it means that scientists who conduct research in situ sometimes have a sort of "local knowledge" that can eventually prove useful, or even indispensable, in achieving research goals.

d) Some Remarks Concerning Research Internationalization and Research Networks in "Resource-Driven" Fields

The spread of the network of international coauthorships is seen across all fields of science, including those characterized as "resource-driven" (Wagner 2005). However, as in other fields, more internationalization does not, in all cases, mean that more countries participate, so more internationalization does not always give rise to geographically distributed scientific networks.[5]

No matter how geographically distributed networks are, resource-driven fields are affected by the *geographical scale of research* in a different way from others, because in this case a wide geographic base might be a sine qua non. Thus, taking into account the dependence upon geographically distant resources, the strong academic competition, and the development of capacities in peripheral countries, we can assume that by increasing the possibilities provided by information and communication technologies (ICTs) to link remote researchers, it is only to be expected that the accumulation of scientific capital in these fields will be closely related to the presence of scientists in international networks and to their capacity to take advantage of the cognitive opportunities provided elsewhere in the globe.

In marine biology, there have been major transformations in the scale of research projects over the last decade. As in the case of the Census of Marine Life, which has been analyzed as an example of big science (Vermeulen 2013), polar marine research evolved toward increasing scale and scope, growing requirements of international coordination, investments, and technology, and strongly networked actors.

Now, let us review our cases. They are the result of fieldwork conducted over eighteen months in CADIC, Ushuaia, analyzing documents, scientific publications, communications between scientists, and sixty in-depth interviews with researchers and PhD students taking part in scientific collaboration activities. For the case discussion, we choose ones that can be seen as "prototypical" of several types of scientific collaboration; we do not wish to be exhaustive in our analysis, but rather show how our main variables can provide a variety of factors that affect the links between local and foreign researchers and groups. The main variables we considered were the driver of the collaboration, the significance and exploitation of local resources, the motivations for collaborating (from both kinds of partners), and the role of techniques and devices.

As stated earlier, our research focused on two fields, geology and marine biology, because these disciplines are good examples of "geographical dependence" and the resource-driven research. Of course, other fields—paleontology, hydrology, and, to an extent, astronomy, among others—may be affected by these features. But in our research "at the end of the world" geology and marine biology display some unusual traits linked to the fact that some scientific material and resources are only be found locally and nowhere else in the world, or offer a privileged site to observe them in that locale. This is clearly different from "laboratory sciences" (Hacking 1992), as we saw in Chapter 5, for the case of molecular biology (even though some research objects, like *Trypanosoma cruzi*, may be viewed as local resources).

2. Scientific Collaborations in Tierra del Fuego

The CADIC was formally created in 1969 as the CONICET's first regional center. However, it was not until the early 1980s that the institute began to operate on a continuous basis, the recruitment of scientists for such a remote region having proved difficult.[6] In fact, a research collaboration agreement with the Spanish National Research Council (CSIC) was explicitly established to encourage Argentine scientists to move to "the South". It can then be said that, unlike most research centers, the beginnings of the CADIC are closely tied to international collaboration.

The CONICET research staff comprises three main categories: scientists belonging to the five-stage research career holding stable positions subject to periodic evaluations, young researchers temporarily benefiting from grants who can later apply for stable positions, and personnel providing technical support. At the time our research was conducted, in the scientific fields covered by this study (which exclude the area of archeology), there were in the CADIC just twenty-three scientists belonging to the CONICET (permanent staff).[7] They were from the following broad scientific areas: geology (comprising five different laboratories or specialties), marine biology (three laboratories), and a third quite heterogonous area comprising terrestrial ecology, forestry, and agronomy research (three laboratories). In addition, there was a laboratory devoted to ecotoxicology and aquatic contamination. Even if public expenditure on scientific research has increased considerably over the last fifteen years (almost doubling), favoring some investment in equipment, laboratories are still poorly equipped. This creates a dependence on other laboratories, particularly those located in the Buenos Aires metropolitan area, but also in La Plata and Córdoba.[8]

a) *General Structure of Scientific Collaborations*

To identify cases of international scientific collaborations, a distinction has been drawn between those implying common research objectives ("Type I" collaborations 82 percent) and other kinds of cooperation activities, which, even if not involving shared research aims, result in copublication ("Type II" collaborations 18 percent). This second type includes various activities, like providing local advice to foreign teams working in the region, preparing review papers in common, or copublishing research results on a common topic obtained independently from different regions. All collaborations with scientists belonging to foreign institutions during the center's short history were considered. Fifty-seven cases were analyzed. Table 9.2 presents a classification of our cases according to three criteria.

Most "Type I" collaborations (38 over 47) depend on data obtained on the basis of the region's "resources" (Type 1.1), whereas some activities (9 of 47) refer to international collaborations unrelated to local research objects (Type 1.2). These figures give us a preliminary idea of "resource-driven" collaborations. The contribution of local scientists must not be understood as only supplying access to resources or empirical data based on them. Quite the contrary; as discussed later, local researchers mostly contribute with their local knowledge, while generally participating in all stages of research. Rather, these figures are indicative of a pattern of the institute's insertion in international networks, as they demonstrate the directionality of empirical data flows.

CADIC research groups link up with scientists from advanced countries through partnerships. Collaborations occur more often with the United States (26 percent) or European countries (53 percent). Taking into account the total number of collaborations, the United States (26 percent), Spain (21 percent), and Chile (16 percent) are the most frequent partners. But if participation is measured in terms of the number of copublished papers or the continuity of partnerships, Germany and the United Kingdom figure among the most significant.[9] Apart from Chile, partnerships with Latin American nations are extremely infrequent. Indeed, collaborations with Chile follow a different pattern: Most of them belong to scientific fields in which the region does not seem to offer geographical advantages, with prevalence of "Type II" collaborations. Furthermore, even when CADIC and Chilean institutions are

Table 9.2 Classification of CADIC Scientific International Collaborations

According to whether there are common research objectives	*According to which country provides access to "resources"*	No.	%
Type I collaborations: Common research objectives	Type 1.1: Argentina (Argentina and other countries: 5) Type 1.2: Abroad (Stays in foreign countries, invited by partner)	38 9	67 16
Type I Collaborations subtotal		47	83
Type II Collaborations: Other activities resulting in copublication		10	17
Type II Collaborations subtotal		10	17
Total		57	100

involved in collaborations with the same European partner on similar or complementary topics, there is no collaboration between them. This suggests peripheral nodes linking only indirectly through central, highly connected vertices. South–South collaborations outside South America are almost nonexistent.

Most collaborations are bilateral, involving the participation of scientists from just two countries. With few exceptions, CADIC has not participated in large-scale programs or projects either promoted, funded, or coordinated by international organizations (be they scientific or intergovernmental). Against such a background, collaborations should be understood as growing out of participants' individual interests in connection with their strategies of scientific capital accumulation and not as an attempt on the part of researchers from scientifically advanced countries to build local capacities.

From the point of view of local researchers, most partners are internationally recognized scientists. When collaborations are based on resources of the region (Type 1.1) most times (66 percent) the topics addressed correspond to the partners' regular research agenda, who are in this way extending to the region research topics in which they already have expertise.

As suggested earlier, the dominance among partners of widely recognized researchers from scientifically advanced countries must be explained by the region's geographical advantages and the potential to address important scientific and global problems that transcend the local scenario.

We asked the scientists to estimate whether the knowledge they intend to produce in region/resources-based collaborations (Type 1.1, total: 38) could be considered as generalizable (and/or having global implications) or, on the contrary, of just local significance. Generalizable knowledge being "high", they considered half of these cases (nineteen) "high" and five of them "intermediate to high". Among these twenty-four "high-impact" cases, the site where the research is performed is "considered to have *unique features* to achieve the research goals".

In terms of differential contributions from local and foreign scientists, the inequality of partners is quite evident (confirming the findings I showed in preceding chapters). By differential contributions, we do not here mean the *only* way in which each partner contributes, but the way that is usually *specific* to each (and, in most cases, could not have been provided by the counterpart). For instance, access to resources and local knowledge is usually a *specific* contribution of local scientists, but this does not exclude participation in all stages of research. In this sense, foreign partners' most frequent contribution is to provide access to technologies that are not available at the local level, a case that may, in some cases, mean that CADIC's scientists could make direct use of their partners' laboratory facilities (52 percent of the total fifty-seven cases). Considering the subgroup of collaborations that depend upon the region's resources (Type 1.1) in 37 percent of cases, Argentine scientists think that research could not have been conducted without some sort of knowledge contribution by the counterpart, and, in 66 percent of cases, because of a lack of funds or unavailability of some technologies. Partners' knowledge contribution refers in almost all cases (except one in the pre-Internet era) to tacit, not explicit, knowledge: mostly laboratory know-how in the case of marine biology and field-work techniques in the case of geology (Shrum 2005).

b) The Role of Techniques and Access to New Technology

Therefore, the lack of development of certain required technologies at the national level is the most decisive factor to explain the need for contributions by foreign scientists. This factor is relevant in all areas but is particularly constraining for local research in geology. Most quantitative technologies used in geology are not available in Argentina,[10] and even if collaborators can contribute with their own laboratory devices and procedures, this does not alter the previous asymmetries, as such technologies cannot be incorporated locally: huge investment and effort would be required, not at the local level (Ushuaia), but at the national level. In such a case, the results of measurements made with techniques provided by partners may enhance local knowledge, but previous inequalities will not only persist, but may even be deepened: In these situations, local scientists are not able to fully capitalize on the results of their own work in terms of scientific recognition.

Instead, certain (less complex or more easily available) laboratory techniques used in marine biology are liable to be incorporated as a result of collaborations. However, even in this more favorable case for local research, these significant but limited efforts are not enough to keep pace with technological change: In one case, a foreign collaborator contributed to the local incorporation of certain molecular biology techniques, but there were no local funds or staff to purchase other sophisticated equipment and acquire the know-how that would enable the type of genomic research that the former partner conducts today. Even if several collaborations have enhanced the local possibility of pursuing certain research objectives, sharp differences in access to technologies are not likely to be overcome on the basis of international collaborations. Differences in equipment required for experimental research are also crucial.

Of course, scientists do not only pursue knowledge production, but promotion in their careers, and copublishing with well-known *core-set* researchers usually contributes to the accumulation of scientific capital at the national level. Consequently, in 80 percent of cases, local scientists feel the collaboration has enhanced their international visibility. An even higher percentage (91 percent) sees the possibility of discussing theoretical or technical issues with their partners as a benefit, and, furthermore, in three-quarters of cases, they believe that this possibility remains open.

3. Foreign Scientists' Motivations to Collaborate with Local Researchers: The Role of Local Resources and Local Knowledge

What are the motivations to collaborate with local scientists from the foreign researchers' viewpoint? As we have seen earlier and in Chapter 8, this issue is directly related to the structure of research problems (Jappe-Heinze 2007) in resource-driven fields (Wagner 2005) that require access to certain regions for the study of uncommon "phenomena", as well as for reconstruction of large-scale processes. But the need to access local resources does not necessarily entail a partnership with local scientists. So, are there other reasons to seek such a partnership?

In approaching this issue, cases are presented to illustrate the role played by the need to access local resources and/or local knowledge and the significance of the resources provided by the region in the context of broader problems. The cases are ordered according to the driver of the cooperation (who has taken the initiative to cooperate). Even if this criterion is insufficient to reveal the nature of the interests that attract foreign scientists, we can think that whenever the collaboration has started at the initiative of the foreign collaborator, the contribution of the local partner was valued "a priori", or even considered a prerequisite, for achieving the expected results (no matter the nature of the contribution, that may vary in different cases).

Out of fifty-six total collaborations reported here, twenty-one correspond to foreign initiatives, another twenty to local researchers' initiatives, eight to what is usually described as "mutual agreement", and the rest to environmental organizations' initiatives. We will look at examples from geology and marine biology, fields that exhibit a remarkable "geographical advantage" in the CADIC. These cases have been chosen because they represent the spectrum of the main different situations that typically lead to the partnership, and not on the basis of their scientific relevance (however it might be measured).[11]

In the comparison of cooperation patterns by scientific field, it is worth noting that foreign initiative is far more common in geology than in marine biology.

The Role of Local Resources and Local Knowledge in Geology

Let us look at some examples of international collaborations in geology in order to shed some light on the type of interests that may steer foreign researchers to partner local ones and on the role played by the need to access local knowledge and resources, which may vary according to the situation.

a) *Foreign Driver I: Local publications on unique or exceptional opportunities in the region to address important research problems that attract the attention of prestigious foreign scientists.*

One of the reasons that explains why foreign initiative is more frequent in geology is the role sometimes played by local researchers—usually through published papers—in calling attention to unique opportunities for research that arise from their own work. This, in turn, attracts scientists from Europe or the United States who seek the partnership. In those cases, local knowledge and resources provided by previous local work are a precondition of the common research, no matter how important the contribution of the foreign scientists might be (and usually is). This situation is illustrated by cases 1 and 2.

Case 1: The research work done over a long period in the James Ross Basin (Antarctica) by an Argentine geologist belonging to the CADIC revealed the almost unique potential of the basin for studying one of the major extinctions occurring in the past: the Cretaceous–Tertiary extinction event, also known as the K–T extinction boundary. The basin contains a complete stratigraphic record (an extremely rare occurrence) that allows comparison with other regions. After the CADIC researcher's publications made this potential clear, two scientists from Caltech and the University of Washington decided to contact him. The exceptional nature of the conditions offered by the basin for testing certain hypotheses about this mass extinction is described by the Caltech collaborator as follows: "It gives us the most incredible ability to go in and get high-resolution data with what was going on with this extinction than anywhere else I've seen, he explained".[12] A partnership was established in which both parts maintained separate but complementary projects, with each collaborating in their partner's project. During the fieldwork, samples were collected for analysis in U.S. laboratories. The local scientist provided his knowledge of the basin, without which the collaborators would not have known where or how to obtain samples, while the U.S. collaborators provided access to several quantitative technologies that were not available in Argentina. By the time we finished our research, this collaboration was still ongoing, with both parties planning to extend their partnership. Due to previous unequal scientific capital, access to funding, and geographical mobility, we can hypothesize that, even if this collaboration contributes to the growth of local researchers' scientific capital, their work still remains "confined" to a region, while the U.S. partners can make wider use of the results in a context of more ambitious research goals concerning the causes of major mass extinctions.

Case 2: After reading a publication by a CADIC geologist (the same as case 1), a prestigious German paleontologist from the Eberhard Karls Universität Tübingen, widely known for his contributions to the study of trace fossils, contacted the Argentine scientist. The local fossil record, together with the German scientist's important collection, helped to address several research problems. By the time we finished our fieldwork, only one of these pieces of research had been completed, resulting in a copublication that had considerable impact. The purpose of the research was to study the particular strategies used by a group of secondary soft, bottom-dwelling organisms (a species of worm) to cope with their earlier loss of mobility. The topic of the research was proposed by the German partner. He provided his wide-ranging expertise in the area, his own important collection, and access to Yale University's collection,[13] while the CADIC scientist provided his local knowledge and collection. Together, these resources were necessary to address the question of whether these organisms' adaptation strategies were genomically programmed. Other collaborations have been planned for the future also involving the use of local collections and/or records as resources, as well as those provided by the German partner, obtained in other parts of the world. The same hypothesis as Case 1 applies here.

b) Foreign Driver II: Geologists from abroad arriving in the
   region with their own research projects but requiring
   local advice.

Of course, in other cases, unlike the examples provided earlier, geologists from Europe or the United States may be aware of the region's potential to address problems in their fields, with little, if any, participation from local scientists. They may, then, travel to Tierra del Fuego with their equipment to conduct field research, sometimes without any previous contact with local scientists. This provides an opportunity to forge links, although sometimes foreign researchers may be reluctant to do so. However, even if foreign geologists arrive in the territory with their own projects, the need for local advice and logistic support may arise.[14] Local scientists may be asked to supply valuable geological information about the region to help identify the most appropriate site for their purposes, such as where to install a GPS station or collect certain samples. Furthermore, local knowledge may occasionally prove crucial in understanding the results of measurements performed in the region, as in the following case, which highlights the importance of contextualized knowledge which can sometimes only be provided by scientists active in the region.

   Case 3: A geologist from the University of Edinburgh, United Kingdom, who had already worked with a local researcher, arrived on the island to apply a relatively new dating technology (cosmogenic nuclides). Efforts to improve measurements using this technique were made simultaneously by the European Union and the United States, which launched important programs with this aim. In Argentina, the visitor applied the technique in the southernmost part of Patagonia (Santa Cruz Province) and attempted to do the same in northern Tierra del Fuego. Local researchers helped him to select the most appropriate site for collecting the samples and participated in the fieldwork. But the results obtained on the island differed greatly from those of Southern Patagonia: To interpret these unexpected results and further determine whether measurements with this technique could be considered reliable in the particular regional conditions of Tierra del Fuego, local knowledge about the region's geological history was essential. This led, in turn, to a sort of "methodological warning" about certain uses of the technique. As stated in the paper coauthored in a mainstream journal in the field, "appreciation of present and past climate regimes can serve as a useful guide to the limit of the probable utility of the technique on old glacial surfaces" (Kaplan et al. 2007).

c) Local Driver: Argentine scientists in search of knowledge and international recognition.

Since the days of the earliest Latin American scientific systems, associating with prestigious groups at the international level has contributed to scientific capital accumulation at the local level (Kreimer 2010a). This trend is currently deepening as a consequence of the well-known growing importance of publications in high-impact international journals, as the most important criterion for bureaucratic evaluation and, hence, for promotion in one's scientific career (Kreimer 2011). It, therefore, comes as no surprise that local scientists are not only willing to accept the invitation of foreign scientists to associate but also develop strategies to achieve this aim, whether guided primarily by the search for knowledge or prestige, or, more frequently, by an indistinguishable combination of both. In our research, we found this kind of strategy particularly in a subfield in which Argentina has very little tradition (quaternary climatic changes) and whose development in the country owes much to a CADIC researcher. The limited tradition at national level, along with the region's outstanding advantages when it comes to tackling problems in this research area,[15] favored a local strategy of scientific capital accumulation closely bound up with international collaborations.

Local geologists guided by an active strategy may deploy several "interessement" devices (Callon 1986) to prove the usefulness of their contribution. This strategy frequently guides the foreign scientists to sites that could be suitable for their research interests. Congresses in the country or region that are capable of attracting prestigious scientists can also be opportunities for such invitations. While the relevance of the contribution of local knowledge may vary in this kind of situation, and sometimes be minor, the act of showing a suitable site (an act that involves geological knowledge about the region) may constitute a precondition for certain projects.

Case 4: A renowned glacial geologist from the Netherlands was studying hydrofractures occurring in glacial settings (also termed "clastic dykes"). To collect samples, he used a specially adapted technique from another scientific field. The Argentine researcher describes the activity of his European colleague when he first met him, as "scouring the globe in search of samples that allowed him to develop his methodology". Though clastic dykes are not infrequent, obtaining the appropriate samples can be difficult. While visiting partners in Argentina, the local scientist offered to guide him to certain sites that could be appropriate for his purposes. One of them, in northern Patagonia, proved suitable, "and now is in the books" he said. Some years later, after studying samples from other countries, the European scientist offered a short series of cases in a coauthored paper establishing, on this basis, "the general characteristics of clastic dykes". They also reconstructed different sets of conditions that may produce them.

The Role of Local Resources and Local Knowledge in Marine Biology

Are marine biologists from scientifically advanced countries primarily motivated by the need to access local resources when associating with local scientists? The answer varies according to the partner and the type of collaboration. Some major actors in sub-Antarctic and Antarctic research are capable of conducting large-scale sampling, so, even if CADIC researchers are invited to take part, local participation does not seem to be required to achieve that aim (case 5). For other scientists, however, obtaining specimens or samples from the Far South costs time and effort, in which case partnerships with local researchers may be welcome (case 6). In an intermediate situation, collaborations in marine biology lasting longer than others do not seem to be exclusively resource-driven, even though this factor probably does count (case 7).

a) Foreign Driver I: Big marine biology in the region; a powerful European institution extending its large-scale surveys to the southern tip of South America and inviting local researchers.

Unlike the case of geology, in which the region attracts the interest of individual, rather than institutional, foreign actors, research in marine biology at the southern tip of South America is characterized by the dominant presence of the Alfred Wegener Institute for Polar and Marine Research (AWI), a leading German institution in polar research. The partnership with this institution connects the CADIC, albeit peripherally, to a strongly internationalized network of actors devoted to Antarctic and sub-Antarctic issues.

Case 5: In a context of strong international networks with high-tech requirements, the German institution has developed a progressive strategy to survey local species, from the southern tip of South America to Antarctica (as part of its pole-to-pole research). The importance of this region lies in the possibility of exploring the ecological, biogeographic, and evolutionary links between the Magellan region and the Antarctic, a topic that has come to be known as "the Magellan–Antarctic connection" and is thought to have implications for trends in global climate change. Furthermore, this region offers what is considered a unique case of ecosystem change and evolution for study.

The German center's major effort with its Latin American counterparts was the LAMPOS (Latin American "Polarstern" Study). Several scientists from South America participated in the survey—among them some CADIC researchers—which was conducted on board the German icebreaker *Polarstern* and focused on the fauna of the Scotia Arc, between the Magellan region and the Antarctic (Arntz & Brey 2003). However, this expedition, which allowed large-scale sampling, can only be understood in the context of previous expeditions by the AWI to the regions located north and south of the Scotia Arc, with Chilean, but above all, European partners.[16] This covered all the regions needed in order to shed light on "the Magellan-Antarctic connection". In this context, it is quite evident that closely networked actors with the capability of performing systemic research had wide access to marine resources in the sub-Antarctic area close to the Tierra del Fuego Archipelago. As a matter of fact, Latin American scientists gained access to certain samples and specimens, but their participation, even if previewed in the LAMPOS expedition, was not required to achieve this aim. Political and strategic decisions about collaborative work with Argentina in the southern tip of South America may possibly be more important reasons for local participation.

b) Foreign Driver II: Trying to get samples from the Far South.

Even if research vessels performing large-scale surveys can help many institutions around the world to gain access to local resources, samples are not available for all scientists interested in them, nor is just any sample obtainable. In this case, foreign scientists can still obtain samples in situ. The procedures usually required to conserve them are very simple, and customs authorization to remove them from the country can be obtained. However, this entails trips to the region, and authorizations take time. So, it is not unusual for local scientists to be asked to collect and send samples to other countries, and foreign scientists may sometimes promise local participation in the research in return. After several disappointments, most marine biologists have decided not to send samples if local participation is uncertain. But, in one way or another, samples may be an object of any negotiations.

Case 6: The case presented here is the only one in which some kind of implicit negotiation over samples gave rise to a collaboration. A CADIC biologist had pursued postgraduate studies overseas on stable isotopes technique as applied to research on the feeding behavior and community dynamics of seabirds. As Antarctic and sub-Antarctic seabirds are highly sensitive to climate change, the relevance of this study goes beyond the local region. Meanwhile, a young scientist from the United States just starting his career had been working in the Antarctic and gathering samples of seabird tissues to be studied in U.S. laboratories using the same technique. The local researcher began to provide some logistic support to her U.S. colleague, and finally, at the latter's initiative, they decided to share a project concerning this topic applied to Antarctic and sub-Antarctic petrels (a type of small seabird). Besides the fact that their research interests were complementary, there were also other benefits for both partners that may have been the primary motivations for collaboration: The Argentine scientist saved his partner's time and effort by obtaining the customs' authorizations needed to take the samples out of the country, while she benefited from sending her own samples to be analyzed in the United States.

c) Mutual Agreement: When access to local resources counts,
but common goals and local knowledge are also important.

The two most relevant international collaborations in marine biology in terms of the number of resulting coauthored papers cannot be reduced to the partner's interest in accessing local resources, however important this factor may be. The following case started before the beginnings of large-scale sampling of the region's sub-Antarctic species, described earlier.[17]

Case 7: During his stay in Europe, one CADIC marine biologist visited a prestigious Scottish scientist working on the ecophysiology of marine fish. The Scottish biologist had been studying fish in the Antarctic, and the encounter raised the opportunity to compare sub-Antarctic and Antarctic notothenioidei, a relevant issue in marine biology, as these fish represent an extremely unusual evolutionary case of adaptive radiation. A project was drawn up together and funded under the European Union's Fourth Framework Program. The CADIC researchers would contribute with their previous knowledge of sub-Antarctic fish and basic laboratory infrastructure. All the foreign participants were considered by CADIC scientists as "top" specialists in the topics covered by the research and made contributions to local incorporation of knowledge, especially with know-how. The partnership with the Scottish scientist lasted several years, with the parties engaging in another common project on the same issue, but with a genomic approach. The collaboration yielded an important number of copublications and contributed to producing knowledge on the topic that later came to be called "the Magellan–Antarctic connection" (see earlier).

d) Environmental Organizations' Initiative: "Charismatic animals" of the region that favor international collaborations.

Studies on sub-Antarctic and Antarctic seabirds became relevant for marine biologists, as mentioned earlier, being especially sensitive to climate change: These animals' foraging behavior, for instance, can be used as an indicator of change in the availability of prey. Beyond cognitive interests, seabirds also attract the attention of environmental organizations. Because these organizations are more successful in raising funds for endangered animals if the species have popular appeal, such species are called "charismatic animals".[18] Penguins, which are only found in the Southern Hemisphere, are a charismatic animal species that garners popular attention. That is why research on endangered penguins in the Southern Ocean can benefit from funds from organizations that interact with local scientists in a variety of ways. The following case is not about a particular case of collaboration, but the different ways in which one of these organizations can participate or act as mediator in international collaborations.

Case 8: The driver for this case was a nongovernmental organization (defining itself as a charity) that conducts and supports scientific research on Antarctic and sub-Antarctic animals to provide a basis for conservation measures. This NGO has contributed to funding CADIC researchers into the foraging ecology of penguin subspecies in Tierra del Fuego, and members of the staff from the organization in European countries coauthored the papers. Though members of the organization who signed as coauthors of these papers have academic degrees, research is not their main activity. The organization draws up its own projects and enlists scientists with the appropriate skills to conduct the research. Scientists from CADIC and from a Chilean institute have been brought together by the organization to work on two different research projects in Chile. CADIC researchers' participation was crucial in both cases, due to their local expertise in certain specific techniques. These partnerships aside, the only other links the NGO recognizes are with nongovernmental organizations. This case addresses the question of whether (and how) environmental organizations are contributing to the growth of international collaborations in science, and to what extent.

4. Closing Remarks

What does this study suggest about the effects of the growing international collaboration in science on the least scientifically developed countries, and particularly about the patterns emerging in "resource-driven" sciences?

Our description of CADIC scientists' international collaborations seems to provide empirical support for the hypothesis, partially suggested by the study of certain international funding mechanisms (Kreimer & Levin 2013) and held by Leydesdorff and Wagner (2008) on the basis of network analysis. The hypothesis runs as follows: Even if the growing international collaboration in science is enhancing the chances of scientists located in the periphery to interact with scientists belonging to the core of the global network, these "core actors" have increased their ability to access, absorb, and make use of research from peripheral contexts even further.

These trends should not be understood as scientists from scientifically advanced countries purposely subordinating local scientists to their own aims, with no regard for their partners' interests.[19] What counts is not cooperativeness as a personal attribute; rather, these trends are the result of the convergence of conditions that seem to amplify the scale and reach in which a specific kind of global research is emerging. While communication and information technologies make it easier to engage in collaborations with scientists working in remote regions than it was in the past, scientists in resource-driven fields are more dependent on the opportunities for research that appear elsewhere in the world than those in other fields. Indeed, in these cases, a wide geographic base may be necessary to address certain problems considered relevant at the global scale.

As capacities are being built in the least scientifically advanced countries, researchers from the periphery can help to draw attention to certain research opportunities that promise to be good investments.[20] When foreign scientists attracted to the region associate with local ones, they are likely to obtain more profits than the latter. Due to the scientific and material capital they already hold, they can profit more from the fieldwork, access more prestigious journals, raise funds from national and international funding agencies, negotiate with firms and regulatory institutions, and so on.

Certainly, in many cases the know-how and/or access to technologies that it provides could not have been supplied by local researchers. Nonetheless, in several cases, previous local work and knowledge constitute a precondition for research. Still, cognitively speaking, foreign partners will often be able to pursue more ambitious—and complex—scientific aims due to their previous positions in international networks and the geographical mobility they enjoy and, therefore, will be able to capitalize on the results of collaborative labors, thereby addressing more general and/or global problems.

Even if their research has some large-scale implications, local scientists will still remain relatively "confined" to the region: not only because they do not usually have the means to pursue the ambitious objectives of their partners (due to lack of funds and/or technological devices, lack of national governments and agencies support), but also because there is still much work to be done to lay the foundations of their scientific knowledge of the region.

Of course, these statements have to be nuanced in various ways. Not all cases or fields in this study share the same features. In geology and paleontology (included here simply as a tool used by geologists), scientists from scientifically advanced countries seem more dependent upon research opportunities revealed by local scientists. Nevertheless, new modalities of collaboration "among unequal partners" coexist with the old forms of knowledge production at the periphery of the globe: as in the past, geological travelers sometimes arrive in the region and disregard any possible contribution by local scientists.

In marine biology, even if the region's resources certainly attract foreign scientists and encourage collaborative work, the transformation of the field into "big science" seems to have left less margin for local scientists to call attention to new research opportunities. They contribute with their specialist knowledge of the region's species, but access to resources appears to be largely dependent on the availability of well-equipped research vessels, which, in the case of Antarctic research, are owned by a handful of major actors performing large-scale sampling in the context of mega-projects and have the capability for conducting systemic research.

This is not the whole story, however. Local scientists are usually willing to collaborate with foreign scientists because the prospect of promotion in their careers has become increasingly dependent upon their publications in high-impact international journals, and the likelihood of publishing in mainstream journals is enhanced by the partnership (Lee & Bozeman 2005; Leydesdorff & Wagner 2008; Wagner et al. 2001, Wagner, Whetsell, & Leydesdorff, 2017). This situation does not drive all researchers to act the same: Some actively seek the partnership, while others will take up any opportunities that arise. Among the former, some may be guided primarily by the search for symbolic profits, funding, travel opportunities, and so on, while, in a few cases, a balance is sought between symbolic and material rewards, and a conscious attempt to acquire knowledge considered useful for the region.

Whatever local scientists' primary motivations, the patterns of collaboration reflect an inequality in terms of the contribution typically specific to each partner. Certainly, some collaborations have allowed access to know-how that would have been harder to acquire if these collaborations had not taken place. But a more frequent reason why local scientists could not have done the research without their partners is lack of funding and local unavailability of certain technologies.

This is, of course, at least in part, a matter of national policy for the less scientifically advanced countries, which are usually—save for certain honorable cases—reluctant to recognize that scientific collaborations among unequal partners are something "good in themselves". Therefore, most nonhegemonic countries tend to encourage all kinds of scientific collaboration—particularly with the most developed countries—without further analysis of the benefits and risks that are usually implicit in these links. Additionally, international collaborations may provide access to funds and equipment and so free local agencies from certain budgetary obligations.

We would like to devote our closing words to the role of large international nongovernmental organizations. Of course, the world of nonprofit organizations is vast and heterogeneous, particularly those devoted to global topics like environmental conservation or, more recently, climate change. They have a wide range of interests (from limited interventions to political negotiations), ideologies (some are very radical and confrontational, while others tend to be more diplomatic), funding (from modest resources to huge funds), partnerships (with other civil organizations, governments, political parties, international agencies, and so on), and capacities (monitoring, research, lobbying, advisement, public actions, and so forth).

One is tempted to think that these kinds of environmental nongovernmental organizations (ENGOs) are usually powerful entities with the potential to recommend or even impose their main topics—including research agendas—on a large swathe of the scientific community. Indeed, this may be true of most large ENGOs and is not necessarily a bad thing; nevertheless, it depends on the specific contexts in which they act. Though stronger ENGOs have the power to intervene at a global level, certain local organizations can also be effective in disputing public topics and even in producing relevant knowledge, as I have shown for a recent case in Chile (Broitman & Kreimer 2017). However, it must be said that local organizations located in developing countries are usually weak and have no availability of funds or research capacities to gear scientific research to socially relevant topics.

Notes

1. *Le Phare du bout du monde* [The Lighthouse at the End of the World] was written by Jules Verne and published in 1905 (the year of Verne's death). It is mainly set on the Isla de los Estados, Tierra del Fuego, not far from Ushuaia. I have taken my title from this emblematic book partly in tribute, partly to convey an idea of the region's remoteness from all the most developed centers.
2. In this perspective, despite the new opportunities for the periphery that such changes entail, nations constituting the core of the global network have increased their ability to "absorb" researchers from outside the core (Leydesdorff & Wagner, 2008). Similar remarks emerge from our analysis of Latin American participation in European Union programs, as we saw in Chapter 8.

3. For an in-depth explanation of the "purification process" see Chapter 3. In a nutshell, it is a procedure that strips a given object of all its local conditions to turn it into a sort of *scientific commodity* that can be moved anywhere to continue the experimentation. Inversely, Knorr-Cetina (1983, p. 119) approached the topic in the early 1980s, stating that:
   "It is clear that measurement instruments are the products of human effort, as are articles, books, and the graphs and print-outs produced. But the source materials with which scientists work are also pre-constructed. Plant and assay rats are specially grown and selectively bred. Most of the substances and chemicals used are purified and are obtained from the industry which serves the science or from other laboratories. The water which runs from a special faucet is sterilized. 'Raw' materials which enter the laboratory are carefully selected and 'prepared' before they are subjected to 'scientific' tests. In short, nowhere in the laboratory do we find the 'nature' or 'reality' which is so crucial to the descriptivist interpretation of inquiry. To the observer from the outside world, the laboratory displays itself as a site of action from which 'nature' is as much as possible excluded rather than included".

4. Efforts related to scientific endeavors in the Antarctic have grown rapidly in recent decades (see Dastidar, 2007), not only because this continent remains largely unexplored from a scientific point of view and so opens up countless possibilities for future research but also because of its significance in shedding light on trends of global climate change. From the point of view of the requirements of marine biology (and other fields of science) to conduct research in the region, specially designed research icebreakers were needed. In the 1980s, cruisers adapted to this end began a systematic search in Antarctica. Today, such facilities are still scarce, "these vessels are few and outdated, thus having to split research efforts between the Arctic and the Southern Ocean" (Lembke-Jene, Biebow, & Thiede, 2011) Partly because of this scarcity, these vessels are frequently shared by countries for common research purposes, a necessity that contributed—though not the only reason—to strengthening international collaboration.

5. While a few resource-driven fields showed widely ranging geographic networks (Wagner, 2005), earth and environmental sciences that study global climate change from a systemic point of view are highly internationalized, with little participation of developing, transition, and emerging countries, because the synoptic observation of global systems requires big investments, computing capacity, and international coordination (Jappe-Heinze, 2007).

6. Ushuaia is more than 3,000 km from Buenos Aires (four to five hours by plane), with an extremely rigorous climate, with an average of 0°C in winter.

7. This figure has risen to over forty scientists today.

8. I characterized this type of dependency in Chapter 7 as research "on the periphery of the periphery", when I analyzed the development of a plasma focus device in Tandil, Province of Buenos Aires.

9. Even if the United States and Spain lead the list of collaborators, they aren't present in all scientific fields; collaborations with scientists from those countries are usual in geology, but neither the United States nor Spain have any considerable weight in marine biology, a field in which partnerships prevail with certain European countries, notably the United Kingdom, France, and Germany.

10. For instance, radiometric dating techniques, stable isotope techniques, magnetostratigraphy, and many others.

11. For the purposes of this chapter, I have excluded international collaborations outside Argentina that are nondependent on data provided by fieldwork done in the region.

12. He added: "Not only that: the rocks are completely unheated, almost unburied, almost no alteration. . . . It's an incredibly exciting sequence to work through. And fossils? You just whack on a few of these things, and boom, outcome ammonites" (statement by Caltech scientist, reproduced in Rejcek, 2010).

13. The curator of Yale's Peabody Museum of Natural History was another collaborator, and the German scientist also worked at Yale, so the Yale collections were made available.

14. Foreign teams are more likely to be there in the research area covered by the "Quaternary Geomorphology" laboratory, so, after long experience, researchers working in this field have developed an explicit strategy to cope with this situation. According to local scientists, they have been disappointed many times by foreign teams to whom they had provided advice and/or logistic support. Yet in such cases, their contribution was not even acknowledged. So now they continue to provide local support and advice on condition they will obtain some benefit in exchange (usually copublication). Not all foreign scientists working on the island will seek this kind of local help (and may even be reluctant to do so) probably because sometimes it is erroneously assumed that there is no useful local scientific expertise in their area of research.

15. Concerning the region's geographical advantages and the types of resources it offers to specialists in this area, southern Patagonia and Tierra del Fuego are said to provide valuable information on certain quaternary climatic changes, not only useful for the region, but also the broader issues of interhemispheric correlations. Such resources have therefore attracted numerous foreign scientists, who have conducted research in the region with or (more often) without participation of local scientists. Regional studies concerning these topics (glaciations, for instance) are, moreover, believed to have implications on the understanding of global climate change.

16. The campaigns were conducted with Chile and Italy in 1994 in the Magellan area, including the eastern entrance to the Beagle Channel (Arntz & Gorny, 1996), with European partners in 2002 to sample deep-sea species (Fütterer et al., 2003), and once again with European partners in a 1988–1989 to study Antarctic ecosystems (European Polarstern Study).

17. Unlike case 5, in which the initiative corresponded to the institution, other collaborations with AWI scientists grew out of individual scientific and personal affinities. This is true of CADIC's prolonged partnership with an AWI marine biologist, not discussed here.

18. A recent paper (Albert, Luque, & Courchamp, 2018) problematizes the use of the term "charismatic" as applied to animal species. Acknowledging that it was diffused by environmental organizations, the authors combine various techniques to establish what the "20 charismatic species" are: "The majority are large exotic, terrestrial mammals. These species were deemed charismatic, mainly because they were regarded as beautiful, impressive, or endangered, although no particular trait was discriminated, and species were heterogeneously associated with most of the traits". Curiously, penguins are not mentioned as "charismatic" in this study, despite attracting much attention from environmental NGOs as an endangered species.

19. In this study, at least some of the foreign partners are described as willing to help, generous, interesting, and such. In two cases where the partners were seen to want to pursue their research according to their interests and at the expense of local wishes, it was decided to interrupt the collaboration.

20. Our focus on global climate change has helped to give legitimacy to topics that in the past deserved less attention.

References

Abir-Am, P. (1992). From multidisciplinary collaboration to transnational objectivity: International space as constitutive of molecular biology. In E. Crawford, T. Shinn, & S. Sörlin (Eds.), *Denationalizing science: The context of international scientific practice, sociology of science yearbook* (Vol. XVI). Dordrecht, The Netherlands: Kluwer.

Abir-Am, P. (2000). *Research schools of molecular biology in the United States, United Kingdom, and France: National traditions or transnational strategies of innovation?* Berkeley, CA: University of California Press.

Abir-Am, P. (2002). The Rockefeller Foundation and the rise of molecular biology. *Nature Reviews Molecular Cell Biology 3*(1), 65–70.

Adams, J. (2012). Collaborations: The rise of research networks. *Nature 490*, 335–336.

Adams, J. (2013). Collaborations: The fourth age of research. *Nature 497*, 557–560.

Adorno, T. (1968). *Introducción a la sociología*. Barcelona: Gedisa.

Agüero, F., Verdún, R. E., Frasch, A. C., & Sánchez, D. O. (2000). A random sequencing approach for the analysis of the trypanosoma cruzi genome: General structure, large gene and repetitive DNA families, and gene discovery. *Genome Research 10*(12).

Albert, C., Luque, G. M., & Courchamp, F. (2018). The twenty most charismatic species. *PLoS One 13*(7), e0199149. https://doi.org/10.1371/journal.pone.0199149

Alexander, P. E., Brito, J. P., Neumann, I., Gionfriddo, M. R., Bero, L., Djulbegovic, B., . . . Guyatt, G. H. (2016). World Health Organization strong recommendations based on low-quality evidence (study quality) are frequent and often inconsistent with GRADE guidance. *Journal of Clinical Epidemiology 72*, 98–106. doi:10.1016/j.jclinepi.2014.10.011

Amadeo, E. (1978). Consejos Nacionales de Ciencia y Tecnología en América Latina: Éxitos y fracasos del primer decenio. *Comercio Exterior 28*(12), 1439–1447.

Amin, S. (1973). *Le developpement inégal*. Paris: Editions de Minuit.

Anderson, M. S. (2011). International research collaborations: Anticipating challenges instead of being surprised. In *The Europa world of learning*. London: Routledge.

Anderson, W. (2009). From subjugated knowledge to conjugated subjects: Science and globalisation, or postcolonial studies of science? *Postcolonial Studies* 12(4), 389–400.

Anderson, W., & Adams, V. (2008). Pramoedya's chickens: Postcolonial studies of technoscience. In E. J. Hackett, O. Amsterdamska, M. Lynch, & J. Wajcman (Eds.), *The handbook of science and technology studies* (pp. 181–204). Cambridge, MA: MIT Press.

Aquino, L. (1921). El profesor Doctor Rodolfo Kraus. *Revista del Círculo Médico Argentino. Centro de Estudios de Medicina*, XX 35–36.

Arntz, W., & Brey, T. (Eds.). (2003). Expedition ANTARKTIS XIX/5 (LAMPOS) of RV Polarstern in 2002. *Ber Polarforsch Meeresforsch 462.*

Arntz, W., & Gorny, M. (Eds.). (1996). Cruise report of the Joint Chilean-German-Italian Magellan "Victor Hensen" Campaign in 1994. *Ber Polarforsch 190.*

Arvanitis, R. (2011). Que des réseaux! Compte rendu de Caroline Wagner: The New invisible college. Science for development. *Revue D'anthropologie Des Connaissances* 5(1), 178–185.

Barclay, C. A., Cerisola, J. A., Lugones, H., Ledesma, O., Lopez Silva, J., & Mouzo, G. (1978). Aspectos farmacológicos y resultados terapéuticos del Benznidazol en el tratamiento de la infección chagásica. *Prensa Médica Argentina* 65, 239–244.

Barnes, B., & Edge, D. (1982). Science as expertise. In B. Barnes & D. Edge (Eds.), *Science in context: Readings in the sociology of science*. Cambridge, MA: The MIT Press.

Barnes, D. (1995). *The making of a social disease: Tuberculosis in nineteenth-century France*. Berkeley, CA: University of California Press.

Basalla, G. (1967). The spread of Western science. A three-stage model describes the introduction of modern science into any non-European nation. *Science* 156(3775), 611–622.

Bassi, S., González, V., & Parisi, G. (2007, December). Computational biology in argentina. *PLoS Computational Biology* 3(12), e257. doi:10.1371/journal.pcbi.0030257

Beaver, D. (2001). Reflections on scientific collaboration (and its study): Past, present, and future. *Scientometrics* 52(3), 365–377.

Beigel, F. (2014). Current tensions and trends in the world scientific system. *Current Sociology* 62(5), 617–625.

Benchimol, J., & Teixeira, L. (1994). *Cobras e largatos & outros bichos. Uma história comparativa dos institutos Butantã e Oswaldo Cruz*. Rio de Janeiro: Fiocruz/Editora da URFJ.

Ben-David, J., & Collins, R. (1966, August). Social, factors in the origins of a new science: The case of psychology. *American Sociological Review 31*(4), 451–465.

Bensaude-Vincent, B. (2001). Materials science and engineering: An artificial discipline about to explode? *History of Recent Science and Technology*. Retrieved from www.library.caltech.edu

Bernaola, O. (2001). *Enrique Gaviola y el Observatorio Astronómico de Córdoba. Su impacto en el desarrollo de la ciencia argentina*. Buenos Aires: Saber y Tiempo.

Bertelotti, M., & Perez Martinez, D. (2008). Gaviotas, ballenas y humanos en conflicto. In *Estado de conservación del mar patagónico y áreas de influencia.* Puerto Madryn. Retrieved from www.marpatagonico.org.

Biagini, G., Escudero, J., Nan, M., & Sánchez, M. (2005). Comentarios a la sentencia de la Corte Suprema de Justicia de la Nación con relación a la obligación del Estado Nacional de suministrar tratamiento antirretroviral a las PPVS. In *Jurisprudencia Argentina* (pp. 52–55). Buenos Aires: Lexis Nexis.

Bibiloni, A. G. (2005). Lecture given at the UNLP on the occasion of the World Year of Physics.

Bijker, W. (1995). *Of bicycles, bakelites, and bulbs: Toward a theory of socio-technical change.* Cambridge, MA: The MIT Press.

Blanckaert, C. (2006). La discipline en perspective: Le système des sciences à l'heure du spécialisme (XIXe–XXe siècle). In J. Boutier, J. C. Passeron, & J. Revel, J. (Eds.), *Qu'est-ce qu'une discipline?* (pp. 117–148). Paris: Éditions de l'EHESS.

Bloor, D. (1976). *Knowledge and social imagery.* London: The University of Chicago Press.

Boekholt, P., Edler, J., Cunningham, P., & Flanagan, K. (2009). *Drivers of international collaboration in research.* Amsterdam: Technopolis Group.

Bonney, K. M., & Engman, D. M. (2015). Autoimmune pathogenesis of Chagas heart disease. *American Journal of Pathology 185*(6), 1537–1547. doi:10.1016/j.ajpath.2014.12.023

Bourdieu, P. (1975). The specificity of the scientific field and the social conditions of the progress of reason. *Social Science Information 14*(6), 19–47.

Bourdieu, P. (1990). Animadversiones in mertonem. In J. Clark & C. Modgil (Eds.), *Robert Merton, consensus and controversy.* London, New York, NY and Philadelphia, PA: Falmer Press.

Bourdieu, P. (1992). *Les règles de l'art, Genèse et structure du champ littéraire.* Paris: Seuil.

Bourdieu, P. (1997). *Les usages sociaux de la science: Pour une sociologie clinique du champ scientifique.* Paris: Editions de l'INRA.

Bourdieu, P. (2001). *Science de la science et réflexivité.* París: Raisons d'agir.

Bozeman, B., & Corley, E. (2004). Scientists' collaboration strategies: Implications for scientific and technical human capital. *Research Policy 33*(4), 599–616.

Bozeman, B., Laredo, P., & Mangematin, V. (2007). Understanding the emergence and deployment of "nano" S&T. *Research Policy 36*(6), 807–812.

Bradley, M. (2007). *North-South research partnerships: Challenges, responses and trends a literature review and annotated bibliography.* Ottawa: Special Initiatives Division, IDRC.

Braun, R. (1989). Crotoxina. *Ciencia Hoy 1*(4), 70–73.

Brawerman, J., & Novick, S. (1982). *Los organismos centrales de Planificación científica y tecnológica en América Latina.* Washington, DC: OEA.

Broitman, C. & Kreimer, P. (2017). Knowledge Production, Mobilization and Standardization in Chile's HidroAyse´n Case. *Minerva*, DOI 10.1007/s11024–017–9335-z

Briceño León, R. (1990). *La Casa Enferma: Sociología de la Enfermedad de Chagas.* Caracas: Fondo Editorial Acta Científica Venezolana.

Buch, A. (2006). *Forma y función de un sujeto moderno: Bernardo Houssay y la fisiología argentina*. Buenos Aires: Editorial de la Universidad Nacional de Quilmes.

Burke, J. G. (1966). *Origins of the science of crystals*. Berkeley, CA: University of California Press.

Cairns, J., Stent, G., & Watson, J. (Eds.). (1966). *Phage and the origins of molecular biology*. New York, NY: Cold Spring Harbor Laboratory of Quantitative Biology.

Callon, M. (1986). Some elements of a sociology of translation: Domestication of the scallops and the fishermen of St Brieuc Bay. In J. Law (Ed.), *Power, action and belief: A new sociology of knowledge?* (pp. 196–223). London: Routledge.

Callon, M., & Latour, B. (1990). *La science telle qu'elle se fait*. Paris: La Découverte.

Cambrosio, A., & Keating, P. (1983). The disciplinary stake: The case of chronobiology. *Social Studies of Science 13*(3), 323–353.

Campomar Foundation. (1985). *The memoirs of the Campomar Foundation (1947–1984)*. Buenos Aires: Campomar Foundation.

Cardoso, F. H., & Faletto, E. (1969). *Dependencia y desarrollo en América Latina*. Buenos Aires: Siglo XXI.

Cereijido, M. (1990). *La nuca de Houssay: La ciencia argentina entre Billiken y el exilio* Buenos Aires: Fondo de Cultura Económica.

Cerisola, J. A. (1977). *Chemotherapy of Chagas' infection in man*. PAHO Scientific Publication, No. 347.

Cerisola, J. A., Da Silva, N. N., Prata, A., Schenone, H., & Rohwedder, R. (1977). Evaluación mediante xenodiagnóstico de la efectividad del nifurtimox en la infección chagásica crónica humana. *Boletín Chileno de Parasitología 32*, 51–62.

Cetto, A. M., & Vessuri, H. (2005). The international scientific cooperation of Latin America and the Caribbean. In *UNESCO world science report*. Paris: UNESCO.

Chatelin, Y., & Arvanitis, R. (1990). Between centers and peripheries: The rise of a new scientific community. *Scientometrics 17*(5–6), 437–452.

Chow-White, P. A., & García-Sancho, M. (2011). Bidirectional shaping and spaces of convergence: Interactions between biology and computing from the first DNA sequencers to global genome databases. *Science, Technology & Human Values 37*(1).

Chubin, D. (1976). The conceptualization of scientific specialties. *The Sociological Quarterly 17*(4).

Ciencia Hoy. (2001). Interview with Lewis Pyenson. Conducted by Miguel de Asúa and José Antonio Pérez Gollán. *11*(65), 58–63.

Coleman, W. (1982). *Death is a social disease: Public health and political economy in early industrial France*. Madison, WI: University of Wisconsin Press.

Collins, H. M. (1975). The seven sexes: A study in the sociology of a phenomenon, or the replication of experiments in physics. *Sociology 9*(2), 205–224.

Collins, H. M. (1981a). The place of the "core set" in modern science: Social contingency with methodological propriety in science. *History of Science, XIX*.

Collins, H. M. (1981b). What is TRASP? The radical programme as a methodological imperative. *Philosophy of the Social Sciences 11*(2), 215.

Collins, H. M. (1983). An empirical relativist programme in the sociology of scientific knowledge. In K. Knorr-Cetina & M. Mulkay (Eds.), *Science observed: Perspectives on the social studies of science*. London: Sage.

Collins, H. M. (1985). *Changing order: Replication and induction in scientific practice*. Chicago: The University of Chicago Press.

Collins, H. M., & Yearley, S. (1992). Epistemological chicken. In A. Pickering (Ed.), *Science as practice and culture*. Chicago: University of Chicago Press.

Community Research and Development Information Service (CORDIS). (2008). *The main objectives of FP7: Specific programmes*. Retrieved December 2009, from http://cordis.europa.eu/fp7/home_en.html

Community Research and Development Information Service (CORDIS). (2009). *Community research and development information service for science, research and development*. Retrieved December 2009, from http://cordis.europa.eu/fp7/understand_en.html

Correa, C. (1996). *Biotecnología: innovación y producción en América Latina*. Buenos Aires: Publicaciones del CBC.

Coutinho, M. (1999). Ninety years of Chagas disease. *Social Studies of Science* 29(4), 519–550.

Craige, W. A. (2015). *The pinboard in practice: A study of method through the case of US telemedicine 1945–1980* (Doctoral thesis). Durham University, Durham.

Creager, A. (1993). Sequences, conformation, information: Biochemists and molecular biologists in the 1950s'. *Journal of the History of Biology 26*(3), 331–360.

Crick, F. (1965). Recent research in molecular biology: Introduction. *British Medical Bulletin 21*, 183.

Cueto, M. (1989). *Excelencia científica en la periferia: actividades científicas e investigación biomédica en el Perú 1890–1950*. Lima: GRADE.

Cueto, M. (1994). Laboratory styles in argentine physiology. *Isis 85*(2), 228–246.

Cueto, M. (1997). Science under adversity: Latin American medical research and American private philanthropy 1920–1960. *Minerva 35*, 233–245.

Cukierman, H. (2007). *Yes, Nos temos Pasteur: Manguinhos, Oswaldo Cruz et a história da ciência no Brasil*. Rio de Janeiro: Relume Fumará-FAPERJ.

Dastidar, P. G. (2007). National and institutional productivity and collaboration in Antarctic science: An analysis of 25 years of journal publications (1980-2004). *Polar Research 26*.

De Chadarevian, S. (2002). *Designs for life: Molecular biology after World War II*. Cambridge: Cambridge University Press.

De Ipola, E. (1997). *Las cosas del creer: Creencia, lazo social y comunidad política*. Buenos Aires: Ariel.

Dear, P. (1994). Book review: *Civilizing mission: Exact sciences and French overseas expansion 1830–1940*, by Lewis Pyenson. *Journal for the History of Astronomy 25*(4), 331–332.

Delaporte, F. (1999). *La maladie de Chagas: histoire d'un fléau continental*. Paris: Ed. Payot et Rivages.

Devan, J., & Tewari, P. S. (2001). Brains abroad. *The McKinsey Quarterly 4*, 1–10.

DNDI. Retrieved from www.dndi.org/cms/public_html/insidecategoryListing.asp?CategoryId=89

Drori, G. (1993). The relationships between science, technology and the economy in lesser developed countries. *Social Studies of Science 23*, 201–215.

Edler, J., & Flanagan, K. (2011). Indicator needs for the internationalisation of science policies. *Research Evaluation 20*(1), 7–17.

Elizari, M. (1999). La Miocardiopatía Chagásica: Perspectiva histórica. *Medicina 59*(Suppl. II), 25–40.

Elizari, M. (2003). Necrológica: En Memoria al Dr. Mauricio B. Rosenbaum. *Revista Argentina de Cardiología 71*(3), 236–238.

El-Sayed, N. M., Myler, P. J., Bartholomeum, D. C., Nilsson, D., Aggarwal, G., Tran, A. N., & Andresson, B. (2005). The genome sequence of trypanosoma cruzi, etiologic agent of Chagas disease. *Science 309*(5733), 409–415.

Epstein, S. (1996). *Impure science AIDS, activism, and the politics of knowledge*. Berkeley and Los Angeles, CA: University of California Press.

Estebanez, M. E. (1996). La creación del Instituto Bacteriológico del Departamento Nacional de Higiene: salud pública, investigación científica y la conformación de una tradición en el campo biomédico. In M. Albornoz, P. Kreimer, & E. Glavich (Eds.), *Ciencia y Sociedad en América Latina* (pp. 427–440). Buenos Aires: Universidad Nacional de Quilmes.

European Commission. (2005). *Impact assessment report on the specific programme International RTD Cooperation Fifth Framework Programme (1998–2002)*. Brussels: Directorate-General for Research, European Commission.

European Commission. (2007a). *FP7 in brief: How to get involved in the EU 7th Framework Programme for Research*. Brussels: European Commission.

European Commission. (2007b). *Livre Vert. L'Espace européen de la recherche: nouvelles perspectives*. Brussels: European Commission.

European Commission. (2008). *Opening to the world: International cooperation in science and technology. Report of the ERA expert group*. Brussels: Directorate-General for Research, European Commission.

European Commission. (2013). *Country and regional scientific production profiles, Luxembourg: Publications office of the European Union*. Brussels: European Commission.

European Commission. (2016). *An analysis of the role and impact of industry participation in the framework programmes*. Brussels: Directorate-General for Research and Innovation, European Commission.

Feld, A. (2009). Estado, comunidad científica y organismos internacionales en la institucionalización de la política científica y tecnológica argentina (1943–1966). In H. Vessuri, P. Kreimer, & A. Arellano (Eds.), *Producción y reflexión sobre Ciencia, Tecnología e Innovación en Iberoamérica*. Caracas: UNESCO-IEASLC.

Feld, A. (2015). *Ciencia y Política(s) en Argentina. 1943–1983*. Buenos Aires: Editorial de la Universidad Nacional de Quilmes.

Feld, A. (2018). El Pensamiento Latinoamericano en Ciencia, Tecnología y Desarrollo (PLACTED) ¿Un pensamiento? ¿Latinoamericano? Una mirada desde el caso argentino. In G. Queluz & T. Brandao (Eds.), *Pensamentos e Identidades em Ciencia e Tecnología no Mundo Iberoamericano*. Curitiba: UTFRP.

Feld, A., & Kreimer, P. (2012). La science en débat en Amérique Latine. Les perspectives "radicales" dans les débuts des années soixante-dix en Argentine. *Revue d'Anthropologie des Connaissances 6*(1).

Ferrari, I. (1997). Towards the physical map of the trypanosoma cruzi nuclear genome: Construction of YAC and BAC libraries of the reference Clone T. Cruzi CL-Brener. *Memorias do Instituto Oswaldo Cruz 92*(6), 843–852.

Ferreira, H. O. (1990). Tratamento da forma indeterminada da doença de Chagas com nifurtimox e benznidazol. *Revista da Sociedade Brasileira de Medicina Tropical 23*, 209–211.

Fornaciari, G., Castagna, M., Viacava, P., Tognetti, A., Bevilacqua, G., & Segura, E. (1992). Chagas' disease in Peruvian Inca mummy. *The Lancet 339*(8785), 128–129.

Fox Keller, E. (2002). *El Siglo Del Gen: Cien Años de Pensamiento Genético.* Barcelona: Península.

Fox Keller, E. (2003). *Making sense of life: Explaining biological development with models, metaphors, and machines.* Cambridge, MA: Harvard University Press.

Fütterer, D., Brandt, A., & Poore, G. (Eds.). (2003). The expeditions ANTARKTIS-XIW3-4 of the research vessel POLARSTERN in 2002 (ANDEEP I and 11: mtarctic benthic deep- sea biodiversity -colonization history and recent community patterns). *Ber Polarforsch Meeresforsch 470.*

Gaillard, J. (1994). North-South research partnership: Is collaboration possible between unequal partners. *Knowledge, Technology & Policy 7*(2), 31–63.

Gaillard, J. (1999). *La coopération scientifique et technique avec les Pays du Sud. Peut-on partager la science ?* Paris: Karthala.

Gaillard, J., & Arvanitis, R. (2013). Science and technology collaboration between Europe and Latin America: Towards a more equal partnership? In J. Gaillard & R. Arvanitis (Eds.), *Research collaboration between Europe and Latin America: Mapping and understanding partnership.* Paris: Editions des Archives Contemporaines.

Gaillard, J., Gaillard, A. M., & Arvanitis, R. (2010). *Mapping and understanding EURO-LAC international cooperation in science and technology.* Paris and Brussels: EULAKS Document, Seventh Framework Programme.

Galison, P. (1997). *Image & logic: A material culture of microphysics.* Chicago: The University of Chicago Press.

Galison, P., & Hevly, B. (1992). *Big science: The growth of large-scale research.* Stanford, CA: Stanford University Press.

Galvão, L. M. C., Nunes, R. M. B., Cançado, J. R., Brener, Z., & Krettli, A. U. (1993). Lytic antibody tire as a jeans of assessing cure after treatment of Chagas disease: A 10 years follow-up study. *Transactions of the Royal Society of Tropical Medicine & Hygiene 87*, 220–223.

García, A., & Teira Serrano, D. (2006). Normas éticas y estadísticas en la justificación de los ensayos clínicos aleatorizados. *Crítica. Revista Hispanoamericana de Filosofía 38*(113).

Gaudillière, J. P. (1993). Molecular biology in the French tradition? Redefining local traditions and disciplinary patterns. *Journal of the History of Biology 26*(3), 473–498.

Gaudillière, J. P. (1996). Molecular biologists, biochemists, and messenger RNA: The birth of a scientific network. *Journal of the History of Biology 29*(3), 417–445.

Gaudillière, J. P. (2001). Making mice and other devices: The dynamics of instrumentation in American Biomedical Research (1930–1960). In T. Shinn & B.

Joerges (Eds.), *Instrumentation between science, state and industry: Sociology of sciences yearbook* (Vols. V & XXII). Dordrecht, The Netherlands, Boston, MA and London: Kluwer Academic Publishers.

Gérard, E., & Cornu, J. F. (2013). Dynamiques de Mobilité étudiante Sud-Nord: Une Approche Par Les Pôles Internationaux de Formation de L'élite Scientifique Mexicaine. *Cahiers Québécois de Démographie 42*(2), 241–272.

Gilbert, W. (1991, January 10). Towards a paradigm shift in biology. *Nature 349*(6305), 99. doi:10.1038/349099a0. Retrieved from www.ncbi.nlm.nih.gov/pubmed/1986314

Gilman, S. (1988). *Disease and representation: Images of illness from madness to AIDS*. Ithaca, NY: Cornell University Press.

Gläser, J., & Laudel, G. (2001). Integrating scientometric indicators into sociological studies: Methodical and methodological problems. *Scientometrics 52*(3), 411–434.

González Casanova, P. (1969). *Sociología de la explotación*. Buenos Aires: Siglo XXI.

Griffiths, E. (2012). *What is a model?* Retrieved August 30, 2018, from https://web.archive.org/web/20120312220527/www.emily-griffiths.postgrad.shef.ac.uk/models.pdf

Gunder Frank, A. (1965). *Capitalismo y subdesarrollo en América Latina*. Buenos Aires: Editorial Signos.

Gusfield, J. (1981). *The culture of public problems: Dinking driving and the symbolic order*. Chicago and London: The University of Chicago Press.

Gusmão, R. (2000). La implicación de los países latinoamericanos en los programas europeos de cooperación CyT con terceros países. *REDES, Revista de Estudios Sociales de la Ciencia 7*(16), 131–163.

Gustin, B. (1973). Charisma, recognition and the motivation of scientists. *American Journal of Sociology 78*, 1119–1134.

Hacking, I. (1992). The self-vindication of the laboratory sciences. In A. Pickering (Ed.), *Science as practice and culture* (pp. 29–64). Chicago: University of Chicago Press.

Hallonsten, O. (2016). Use and productivity of contemporary, multidisciplinary big science. *Research Evaluation 25*(4), 486–495. https://doi.org/10.1093/reseval/rvw019

Halperin Donghi, T. (1962). *Historia de la Universidad de Buenos Aires*. Buenos Aires: EUDEBA.

Haraway, D. (1991). A cyborg manifesto: Science, technology, and socialist-feminism in the late twentieth century. In D. Haraway (Ed.), *Simians, cyborgs and women: The reinvention of nature*. New York, NY: Routledge.

Harding, S. (2008). *Science from below*. Durham, NC and London: Duke University Press.

Harrison, M. (1995). Lewis Pyenson, civilizing mission: Exact sciences and French overseas expansion 1830–1940. *The British Journal for the History of Science 28*(1), 119–120.

Headrick, D. (1995). Review of Pyenson 1993. *The American Historical Review 100*(4), 1260.

Heilbron, J. (2004). A regime of disciplines: Toward a historical sociology of disciplinary knowledge. In C. Camic & H. Joas (Eds.), *The dialogical turn: New roles for sociology in the post disciplinary age*. New York, NY and Oxford: Rowman & Littlefield.

Helden, V., & Hankins, T. (1994). Introduction: Instruments in the history of science. *Osiris*, second series 9, 1–6.

Hilgartner, S. (1995). The human genome project. In S. Jasanoff, G. Markle, J. Peterson, & T. Pinch (Eds.), *Handbook of science and technology studies* (pp. 302–315). Thousand Oaks, CA: Sage.

Hilgartner, S. (2004). Making maps and making social order: Governing American Genome Centers 1988–93. In J. P. Gaudilliere & H. J. Rheinberger (Eds.), *From molecular genetics to genomics: The mapping cultures of twentieth-century genetics* (pp. 113–128). New York, NY: Routledge.

Hill, S. (1986). The hidden agenda of science studies for developing countries. *Science & Technology Studies* 4(3/4), 29–32.

Hine, C. (2006, April 1). Databases as scientific instruments and their role in the ordering of scientific work. *Social Studies of Science* 36(2), 269–298. doi:10.1177/0306312706054047

Hontebeyrie-Joskowicz, M., & Minoprio, P. (1991). Trypanosoma cruzi versus the host immune system. *Research in Immunology* 142(2), 125.

Hurtado de Mendoza, D., & Vara, A. M. (2007). Winding roads to big science: Experimental Physics in Argentina and Brazil. *Science Technology & Society* 12(1).

INDIECH. (1995). *Memoria*. Buenos Aires: Institute for Diagnosis and Research of Chagas Disease.

Inkster, I. (1985). Scientific enterprise and the colonial "model": Observations on Australian experience in historical context. *Social Studies of Science* 15(4), 677–704.

Institut Pasteur. (1987). *1887–1987–2087: Un Nouveau Siècle*. Paris: Editions de l'Institut Pasteur.

Jappe-Heinze, L. A. (2007). *Knowledge about the spaceship earth: A sociological perspective on capacity development* (Dissertation eingereicht zur Erlangung des Gradeseines Doktors der Philosophie (Dr. phil.) der Fakultät für Soziologie Universität Bielefeld).

Jasanoff, S. (1995). Procedural choices in regulatory science. *Technology in Society* 17(3), 279–293.

Jasanoff, S. (1998). *The fifth branch: Science advisers as policymakers*. Cambridge, MA: Harvard University Press.

Jasanoff, S. (2004). The idiom of coproduction. In S. Jasanoff (Ed.), *States of knowledge: The co-production of science and social order*. New York, NY: Routledge.

Joerges, B., & Shinn, T. (Eds.). (2001). *Instrumentation between science, state and industry*. Dordrecht, The Netherlands and London: Kluwer Academic Publishers.

Joly, P. B., Rip, A., & Callon, M. (2010). Reinventing innovation. In M. Arentsen, W. van Rossum, & B. Steenge (Eds.), *Governance of innovation*. Cheltenham: Edward Elgar.

Kaplan, M. R., Coronato, A., Hulton, N. R. J., Rabassa, J. O., Kubik, P. W., & Freemand, S. P. H. T. (2007). Cosmogenic nuclide measurements in southernmost South America and implications for landscape change. *Geomorphology* 87(4).

Katz, J. S., & Bercovich, N. (1990). *Biotecnología y economía política: estudios del caso argentine*. Buenos Aires: Centro Editor de América Latina.

Katz, J. S., & Martin, B. R. (1997). What is research collaboration? *Research Policy 26*, 1–18.

Kaufmann, A. (2004). Mapping the human genome at Genethon laboratory—the French Muscular Dystrophy Association and the politics of the gene. In J.-P. Gaudillière & H.-J. Rheinberger (Eds.), *From molecular genetics to genomics: The mapping cultures of twentieth-century genetics*. London: Routledge.

Kay, L. E. (1993). *The molecular vision of life: Caltech, the Rockefeller Foundation and the new biology*. New York: Oxford University Press.

Kay, L. E. (1996). The tools of the discipline: Biochemists and molecular biologists. *Journal of the History of Biology 29*(3), 446–447.

Kendrew, J. (1967). How molecular biology started. *Scientific American 216*(1967), 141–143.

Klein, E. (2011). *Le small bang des nanotechnologies*. Paris: Odile Jacob.

Knorr Cetina, K. (1982). Scientific communities or transepistemic arenas of research? A critique of cuasi-economic models of research. *Social Studies of Science 12*(1), 101–130.

Knorr Cetina, K. (1983). The ethnographic study of scientific work: Towards a constructivist interpretation of science. In K. Knorr-Cetina & M. Mulkay (Eds.), *Science observed: Perspectives on the social study of science* (pp. 115–140). London: Sage.

Knorr Cetina, K. (1995). Laboratory studies, the cultural approach to the study of science. In S. Jasanoff, G. Markle, J. Petersen, & T. Pinch. *Handbook of Science and Technology Studies*. London, New Delhi and Thousand Oaks, CA: Sage.

Knorr-Cetina, K. (1999). *Epistemic cultures: How sciences make knowledge*. Cambridge, MA: Harvard University Press.

Kreimer, P. (1996). Science and politics in Latin America: The old and the new context in Argentina. *Science, Technology and Society 2*(1), 290–315.

Kreimer, P. (1998). Understanding scientific research on the periphery: Towards a new sociological approach? *EASST Review 17*(4), 13–21.

Kreimer, P. (1999a). *De probetas, computadoras y ratones: La construcción de una mirada sociológica sobre la ciencia*. Buenos Aires: Editorial de la Universidad Nacional de Quilmes.

Kreimer, P. (1999b). *L'universel et le contexte dans la recherche scientifique*. Lille: Presses Universitaires du Septentrion.

Kreimer, P. (2000). Ciencia y periferia, una lectura sociológica. In M. Monserrat (Ed.), *La ciencia argentina entre siglos*. Buenos Aires: Manantial.

Kreimer, P. (2006). ¿Dependientes o integrados? La ciencia latinoamericana y la división internacional del trabajo. *Nómadas 24*, 199–212.

Kreimer, P. (2010a). La recherche en Argentine: entre l'isolement et la Dépendance. *Cahiers de la recherche sur l'éducation et les savoirs 9*.

Kreimer, P. (2010b). Las tensiones de Varsavsky. In O. Varsavsky (Ed.), *Ciencia, política y cientificismo*. Buenos Aires: Capital Intelectual.

Kreimer, P. (2010c). Institucionalización de la investigación científica en la Argentina: de la internacionalización a la división internacional del trabajo científico. In G. Lugones & J. Flores (Eds.), *Intérpretes e interpretaciones de la Argentina en el Bicentenario.* Buenos Aires: Editorial de la Universidad Nacional de Quilmes.

Kreimer, P. (2010d). *Ciencia y Periferia. Nacimiento, muerte y resurrección de la biología molecular en la Argentina. Aspectos sociales, políticos y cognitivos.* Buenos Aires: EUDEBA.

Kreimer, P. (2011). La evaluación de la actividad científica: desde la indagación sociológica a la burocratización. Dilemas actuales. *Propuesta educativa*, (36), 59–77.

Kreimer, P. (2012). Délocalisation des savoirs en Amérique latine: le rôle des réseaux scientifiques. *Pouvoirs Locaux 94*(3), 26–30.

Kreimer, P. (2015). Los mitos de la ciencia: desventuras en las prácticas científicas, los estudios sobre la ciencia y las políticas científicas. *Nomadas*, (42), 32–51.

Kreimer, P. (2016a). Contra viento y marea en la ciencia de la modernidad periférica: niveles de análisis, conceptos y métodos. In P. Kreimer (Ed.), *Contra viento y marea: Emergencia y desarrollo de campos científicos en la periferia.* Buenos Aires: CLACSO.

Kreimer, P. (2016b). Co-producing social problems and scientific knowledge: Chagas disease and the dynamics of research fields in Latin America. In M. Merz & P. Sormani (Eds.), *The sociology of science yearbook 29. The local configuration of new research fields. On regional and national diversity.* Dordrecht, The Netherlands: Springer.

Kreimer, P. (2017). An unrequited love: Social sciences and STS. *Revue d'anthropologie des connaissances 11*(2), ce–cx. doi:10.3917/rac.035.0186

Kreimer, P., & Corvalán, D. (2010). 20 años no es nada. Conocimiento científico, producción de medicamentos y necesidades sociales. *Desarrollo Económico 49*(193).

Kreimer, P., & Levin, L. (2013). Scientific cooperation between the European Union and Latin American countries: Framework programmes 6 and 7. In J. Gaillard & R. Arvanitis (Eds.), *Research collaborations between Europe and Latin America: Mapping and understanding partnership.* Paris: Editions des Archives Contemporaines.

Kreimer, P., & Lugones, M. (2003). Pioneers and victims: The birth and death of Argentina's first molecular biology laboratory. *Minerva 41*, 47. https://doi.org/10.1023/A:1022209922391

Kreimer, P., & Meyer, J. B. (2008). Equality in the networks? Some are more equal than others. International scientific cooperation: An approach from Latin America. In H. Vessuri & U. Teichler (Eds.), *Universities as centers of research and knowledge creation: An endangered species?* Rotterdam: Sense Publishers.

Kreimer, P., & Thomas, H. (2005). Production des connaissances dans la science périphérique: l'hypothèse CANA en Argentine. In J. B. Meyer & M. Carton (Eds.), *La société des savoirs: Trompe-l'œil ou perspectives?* Paris: L'Harmattan.

Kreimer, P., & Vessuri, H. (2017). Latin American science, technology, and society: A historical and reflexive approach. *Tapuya: Latin American Science, Technology and Society 1*(1), 17–37. doi:10.1080/25729861.2017.1368622

Kreimer, P., Vessuri, H., Velho, L., & Arellano, A. (2014). El estudio social de la ciencia y la tecnología en América Latina: miradas, logros y desafíos. In P. Kreimer, H. Vessuri, L. Velho, & A. Arellano (Eds.), *Perspectivas latinoamericanas en el estudio social de la ciencia, la tecnología y la sociedad*. México DF: Siglo XXI.

Kreimer, P., & Zabala, J. P. (2006). ¿Qué conocimiento y para quién? Problemas sociales, producción y uso social de conocimientos científicos sobre la enfermedad de Chagas en Argentina. *REDES, Revista de Estudios Sociales de la Ciencia* 12(23), 49–78.

Kreimer, P., & Zabala, J. P. (2007). Chagas disease in Argentina: Reciprocal construction of social and scientific problems. *Science Technology and Society*, (12), 49–72.

Kreimer, P., & Zabala, J. P. (2009). Quelle connaissance et pour qui? Problèmes sociaux, production et usage social de connaissances scientifiques sur la maladie de Chagas en Argentine. *Revue d'Anthropologie des Connaissances* 3(5), 413–439.

Kropf, S., Azevedo, N., & Ferreira, L. (2003). Biomedical research and public health in Brazil: The case of Chagas' disease. *Social History of Medicine* 16(1), 111–129.

Kuhn, T. (1970). *The structure of scientific revolutions*. Chicago: The University of Chicago Press.

Lafuente, A., & Sala Catalá, J. (1992). Ciencia y mundo colonial: el contexto iberoamericano. In A. Lafuente & J. Sala Catalá (Eds.), *Ciencia colonial en América*. Madrid: Alianza.

Latour, B. (1983). Give me a laboratory and I will raise the world. In K. Knorr Cetina & M. Mulkay (Eds.), *Science observed* (pp. 141–170). London: Sage.

Latour, B. (1989). *Science en action*. Paris: La Découverte.

Latour, B. (1991). *Nous n'avons jamais été modernes: Essai d'anthropologie symétrique*. Paris: La Découverte.

Latour, B. (1999). *Politiques de la nature*. Paris: La Découverte.

Latour, B. (2000). On the partial existence of existing and nonexisting objects. In L. Daston & J. Renn (Eds.), *The coming into being and the passing away of scientific objects*. Chicago: The University of Chicago Press.

Latour, B. (2005). *Reassembling the social*. New York, NY: Oxford University Press.

Latour, B., & Woolgar, S. (1979). *Laboratory life: The construction of scientific facts*. Beverly Hills: Sage.

Law, J. (2004). *After method: Mess in social science research*. New York, NY: Routledge.

Law, J. (2006). Interview with John Law. *Redes, Revista de Estudios Sociales de la Ciencia* 11(24).

Law, J., & Lin, W. (2017). Provincializing STS: Postcoloniality, symmetry, and method. *East Asian Science, Technology and Society* 11(2), 211–227.

Lee, S., & Bozeman, B. (2005). The impact of research collaboration on scientific productivity. *Social Studies of Science* 35(5), 673–702.

Leloir, L. F. (1982). Cincuenta años con la ciencia. Allá lejos y hace tiempo. *Acta Bioquímica Latinoamericana, XX*(3), 301–331.

Lemaine, G., Weingart, P., MacLeod, R., & Mulkay, M. (Eds.). (1976). *Perspectives on the emergence of scientific disciplines*. Paris, Chicago and The Hague, The Netherlands: Mouton.

Lembke-Jene, L., Biebow, N., & Thiede, J. (2011). The European research ice-breaker aurora borealis, conceptual design study—summary report. *Reports on Polar and Marine Research 637.*

Lenoir, T. (1993). The discipline of nature and the nature of disciplines. In E. Messer-Davidow, D. R. Shumway, & D. Sylvan (Eds.), *Knowledges: Historical and critical studies in disciplinarity* (pp. 70–102). Charlottesville, VA: University Press of Virginia.

Lenoir, T. (1997). *Instituting science: The cultural production of scientific disciplines.* Stanford, CA: Stanford University Press.

Lenoir, T. (1999). Shaping biomedicine as an information science. In M. E. Bowden, T. B. Hahn, & R. V. Williams (Eds.), *Proceedings of the 1998 conference on the history and heritage of science information systems* (pp. 27–45). Medford, NJ: Information Today.

Lepenies, W., & Weingart, P. (1983). Introduction. In L. Graham, W. Lepenies, & P. Weingart (Eds.), *Functions and uses of disciplinary histories.* Dordrecht, The Netherlands: Reidel.

Levin, L., Jensen, P., & Kreimer, P. (2016). Does size matter? The multipolar international landscape of nanoscience. *PLoS One 11*(12), e0166914. https://doi.org/10.1371/journal.pone.0166914

Levin, M. J., Rossi, R., Levitus, G., Mesri, E., Bonnefoy, S., Kerner, N., & Hontebeyrie-Joskowicz, M. (1990). The cloned C-terminal region of a Trypanosoma cruzi P ribosomal protein harbors two antigenic determinants. *Immunology Letters 24*, 69–74.

Leydesdorff, L., & Wagner, C. (2008). International collaboration in science and the formation of a core groupe. *Journal of Informetrics 2*(4), 317–325.

Lorenzano, C. (1994). *Por los caminos de Leloir: Estructura y desarrollo de una investigación Nobel.* Buenos Aires: Biblos.

Losego, P., & Arvanitis, R. (2008). La science dans les pays non hégémoniques. *Revue d'Anthropologie des Connaissances 2*(3).

Löwy, I. (2001). *Virus, moustiques et modernité: La fièvre jaune au Brésil, entre science et politique.* Paris: Editions des Archives Contemporaines.

Lwoff, A. (1966). The prophage and I. In J. Cairns, G. Stent, & J. Watson (Eds.), *Phage and the origins of molecular biology* (pp. 88–99). New York, NY: Cold Spring Harbor Laboratory of Quantitative Biology.

Lwoff, A. (1981). *Jeux et combats.* Paris: Fayard.

MacKenzie, D. (1981). *Statistics in Britain 1865–1930: The social construction of scientific knowledge.* Edinburgh: Edinburgh University Press.

MacLeod, R. (1980). On visiting the "moving metropolis": Reflections on the architecture of imperial science. *Historical Records of Australian Science 5*(3), 1–16.

MacLeod, R. (2000). Nature and empire: Science and the colonial enterprise. Introduction.*Osiris 15.*

Manzur, R., & Barbieri, G. (2002). *Enfermedad de Chagas Crónica: aspectos de controversia sobre tratamiento etiológico.* Paper presented at XXth Simposyum on Chagas Disease, National Cardiological Congress, Buenos Aires, April 29–May 2.

Marcovich, A., & Shinn, T. (2011). Where is disciplinarity going? Meeting on the borderland. *Social Science Information 50*(3–4), 582–606.

Marí, M. (1982). *Evolución de las concepciones de política y planificación de ciencia y tecnología.* Washington, DC: OAS.

Marini, R. M. (1973). *Dialéctica de la dependencia.* Buenos Aires and Bogotá: CLACSO-Siglo del Hombre Editores.

Mazza, S. (1949). La enfermedad de Chagas en la República Argentina. *Memorias do Instituto Oswaldo Cruz 47,* 1–2.

McKusick, V. A., & Ruddle, F. H. (1987). A new discipline, a new name, a new journal. *Genomics 1*(1), 1–2. doi:10.1016/0888-7543(87)90098-X

Merton, R. K. (1938). Science, technology and society in seventeenth century England. *Osiris 4,* 360–632.

Merton, R. K. (1968). *Social theory and social structure.* New York, NY: The Free Press.

Meyer, M. (2007). What do we know about innovation in nanotechnology? Some propositions about an emerging field between hype and path-dependency. *Scientometrics 70*(3), 779–810.

Mignolo, W. (2002). The geopolitics of knowledge and the colonial difference. *The South Atlantic Quarterly 101*(1), 57–96.

Ministry of Health and Social Action. (1983). *Resolución Secretaría de Programas de Salud.* Buenos Aires: Ministry of Health and Social Action.

Ministry of Health and Social Action. (1998). *Manual para la atención de pacientes infectados con Chagas.* Buenos Aires: Ministry of Health and Social Action.

Ministry of Health and Social Action. (2005). *Manual para la atención de pacientes infectados con Tripanosoma Cruzi.* Buenos Aires: Ministry of Health and Social Action.

Morange, M. (1994). *Histoire de la biologie moléculaire.* Paris: La Découverte.

Moravcsik, M. (1985). Science in the developing countries: An unexplored and fruitful area for research. *4S Review 3*(3), 2–13.

Muldur, U., Corvers, F., Delanghe, H., Dratwa, J., Heimberger, D., Sloan, B., & Vanslembrouck, S. (2006). *A new deal for an effective European research policy. The Design and impacts of the 7th framework programme.* Dordrecht, The Netherlands: Springer.

Mulkay, M. (1976). The mediating role of the scientific elite. *Social Studies of Science 6*(3/4).

Mulkay, M. (1991). *Sociology of science: A sociological pilgrimage.* London: Open University Press.

Mullins, N. (1972). The development of a scientific speciality: The Phage group and the origins of molecular biology. *Minerva 10*(1), 51–81.

National Science Board. (2008). *International science and engineering partnerships: A priority for U.S. Foreign Policy and our nation's innovation enterprise.* Washington, DC: The National Academies.

National Science Foundation. (2008). *Research and development: Essential foundation for US competitiveness in a global economy.* Washington, DC: NSF.

National Secretariat of Science and Technology (SECyT). (1989). *Aportes para la memoria.* Buenos Aires: SECyT.

Newman, M. (2001). The structure of scientific collaboration networks. *Proceedings of the National Academy of Sciences USA 98,* 404–409.

OECD. (2002). *International mobility of the highly skilled.* http://dx.doi.org/10.1787/9789264196087-en. Retrieved January 2013, from www.oecd.org/sti/inno/internationalmobilityofthehighlyskilled.htm

OECD. (n.d.). *Main science and technology indicators*. Retrieved June 2016, from www.oecd.org/sti/msti.htm

Olby, R. (1994). *The path to the double helix: The discovery of DNA*. New York: Dover.

Ortiz, E. (2010). The emergence of theoretical physics in Argentina, Mathematics, mathematical physics and theoretical physics 1900–1950. *Proceedings of Science* (HRMS) 030. https://pos.sissa.it/109/030/pdf

Oszlak, O., & O'Donnell, G. (1995 [1981]). Estado y políticas estatales en América Latina. Hacia una estrategia de investigación. *Redes, Revista de Estudios Sociales de la Ciencia* 2(4).

Oteiza, E. (1992). *La política científica y tecnológica en Argentina*. Buenos Aires: CEAL.

PAHO/WHO. (1998). *Tratamiento Etiológico de la Enfermedad de Chagas, Conclusiones de una consulta Técnica*. Rio de Janeiro: Panamerican Health Organization.

Persson, O., Glänzel, W., & Danell, R. (2004). Inflationary bibliometric values: The role of scientific collaboration and the need for relative indicators in evaluative studies. *Scientometrics* 60(3), 421–432.

Pestre, D. (1992). Les physiciens dans les sociétés occidentales de après-guerre Une mutation des pratiques techniques et des comportements sociaux et culturels. *Revue Histoire moderne et contemporaine* 39(1).

Pestre, D. (1995). Pour une histoire sociale et culturelle des sciences. Nouvelles définitions, nouveaux objets, nouvelles pratiques. *Annales. Histoire, Sciences Sociales* 50(3). 487–522.

Pestre, D. (2003a). *Science, argent et politique*. Paris: Éditions de l'INRA.

Pestre, D. (2003b). Regimes of knowledge production in society: Towards a more political and social reading. *Minerva* 41(3). doi:10.1023/A:1025553311412

Petitjean, P. (1996). Introduction. In R. Waast (Ed.), *Les Sciences hors d'Occident au XX siècle*. Vol. 2: *Les Sciences Coloniales, Figures et Institutions*. Paris: ORSTOM.

Petitjean, P., Jami, C., & Moulin, A. M. (Eds.). (1992). *Sciences and empires: Historical studies about scientific development and European expansion*. Dordrecht, The Netherlands, Boston, MA and London: Kluwer.

Pickering, A. (1995). *The mangle of practice time, agency, and science*. Chicago: University of Chicago Press.

Pickering, A. (1999). *Constructing quarks: A sociological history of particle physics*. Chicago: University of Chicago Press.

Pirosky, I. (1986). *1957–1962. Progreso y Destrucción de Instituto Nacional de Microbiología*. Buenos Aires: EUDEBA.

Poncet, C. (2001). *De la connaissance académique à l'innovation industrielle dans les sciences du vivant: Essai d'une typologie organisationnelle dans le processus d'industrialisation des connaissances*. Paper presented to the Congress Travail et Mondialisation (IRD), Paris, December 6 & 7.

Prebisch, R. (1951). Crecimiento, desequilibrio y disparidades: interpretación del proceso de desarrollo. In *Estudio Económico de América Latina* 1949. Santiago de Chile: ECLAC.

Price, D. (1963). *Little science, big science*. New York, NY: Columbia University Press.

Pyenson, L. R. (1985). *Cultural imperialism and exact sciences: German expansion overseas 1900–1930*. Studies in History and Culture (Vol. 1). New York, NY, Bern, and Frankfurt: Peter Lang.

Pyenson, L. R. (1993). *Civilizing mission: Exact sciences and French overseas expansion 1830–1940*. New York, NY: Johns Hopkins University Press.

Rabeharisoa, V., & Callon, M. (2008). The involvement of patients' associations in research. *International Social Science Journal 54*(171), 57–63. doi. org/10.1111/1468-2451.00359

Rabinowicz, P. D. (2001). Genomics in Latin America: Reaching the frontiers. *Genome Research 11*, 319–322. doi:10.1101/gr.179501.

Raj, K. (2000). Colonial encounters and the forging of new knowledge and national identities: Great Britain and India 1760–1850. *Osiris 2nd Series, Nature and Empire: Science and the Colonial Enterprise 15*, 119–134.

Raj, K. (2016). Go-betweens, travelers, and cultural translators. In B. Lightman (Ed.), *A companion to the history of science*. Chichester: John Wiley & Sons.

Reed, B. C. (2014). *The history and science of the Manhattan project, undergraduate lecture notes in physics*. Berlin, Heidelberg: Springer-Verlag.

Rejcek, P. (2010). What killed the dinosaurs? Scientists believe asteroid theory may not tell the whole story of KT extinction. *The Antarctic Sun*, News about the USAP, the ice and the people, United States Antarctic Program. Retrieved March 2012, http://antarcticsun.usap.gov/science/contentHandler. cfm?id=1534

Rieznik, M. (2008). *Historia de la Astronomía en la Argentina: Los observatorios de Córdoba y de La Plata (1871–1935)* (Doctoral Thesis). Facultad de Filosofía y Letras, Universidad de Buenos Aires, Buenos Aires.

Rodriguez Medina, L. (2014). *Centers and peripheries in knowledge production*. New York, NY and London: Routledge.

Romero, F. (2004). *Manuel Elkin Patarroyo: un scientifique mondial. Inventeur du vaccine de synthèse de la malaria*. Paris: L'Harmattan.

Rosenbaum, M. B., & Alvarez, J. (1953, October). Miocarditis crónica chagásica en la Provincia de Buenos Aires. *El Día Médico* 1898–1901.

Rosenbaum, M. B., & Cerisola, J. A. (1957). Encuesta sobre la enfermedad de Chagas en el norte de Córdoba y sur de Santiago del Estero. *La Prensa Médica Argentina 44*(35), 2713–2727.

Rosenbaum, M. B., & Cerisola, J. A. (1958). Encuesta sobre la enfermedad de Chagas en la Provincia de La Rioja. *La Prensa Médica Argentina 45*(10), 1013–1026.

Rostow, W. W. (1962). *The stages of economic growth* (pp. 2, 38, 59). London: Cambridge University Press.

Rouquié, A. (1998). *Amérique Latine: Introduction à l'extrème occident*. Paris: Editions du Seuil.

Saldaña, J. J. (1992). Acerca de la historia de la ciencia nacional. *Cuadernos de Quipu*, (4), 9–54.

Saldaña, J. J. (1996). *Historia social de las ciencias en América Latina*. Mexico DF: UNAM.

Salomon, J. J. (1994). Modern science and technology. In J. J. Salomon, F. Sagasti, & C. Sachs (Eds.), *The uncertain quest. Science: Technology and development*. Tokyo: The United Nations University Press.

Sanmartino, M. (2005). Enfermedad de Chagas: concepciones de los habitantes de un área endémica. In M. Abramzón, L. Findling, A. M. Mendes Diz, & P. Di Leo (Eds.), *VI Jornadas Nacionales de Debate Interdiscipliario de Salud y Población*. Buenos Aires: Instituto Gino Germani.

Sanmartino, M., & Crocco, L. (2000). Conocimientos sobre la enfermedad de Chagas y factores de riesgo en comunidades diferentes de la Argentina. *Revista Panamericana de Salud Pública* (OPS), 7(3).

Santesmases, M. J., & Muñoz, E. (1997). *Establecimiento de la bioquímica y de la biología molecular en España (1940–1970)*. Madrid: Fundación Ramón Areces, Consejo Superior de Investigaciones Científicas.

Sarlo, B. (1988). *Una modernidad periférica: Buenos Aires 1920 y 1930*. Buenos Aires: Nueva Visión.

Sasson, A. (1988). *Biotechnologies and development*. París: Unesco/CTA.

Schapachnik, E. (2002). El tratamiento antiparasitario en la enfermedad de chagas, ¿debe darse a todos o no? A favor. *Revista Argentina de Cardiología 70*(5).

Sebastián, J. S. (2007). Conocimiento, cooperación y desarrollo. *Revista CTS 3*(8), 195–208.

Secord, J. A. (2004). Knowledge in transit. *Isis*, (4), 654–672.

Segura, E. (2002). El control de la Enfermedad de Chagas en la República Argentina. In PAHO (Ed.), *El control de la Enfermedad de Chagas en los países del cono sur de América. Historia de una iniciativa internacional 1991/2001* (pp. 45–96). Washington, DC: PAHO.

Shapin, S., & Schaffer, S. (1985): *Leviathan and the air-pump: Hobbes, Boyle, and the experimental life*. Princeton, NJ: Princeton University Press.

Shils, E. (1961). Centre and periphery. In *The logic of personal knowledge: Essays presented to Michael Polanyi* (pp. 117–130). London: Routledge & Kegan Paul.

Shils, E. (1975). *Center and periphery: Essays in macrosociology*. Chicago: University of Chicago Press.

Shils, E. (1988). Center and periphery: An idea and its career 1935–1987. In L. Greenfeld & M. Martin (Eds.), *Center: Ideas and institutions*. Chicago: The University of Chicago Press.

Shinn, T. (1988). Hiérarchies des chercheurs et formes des recherches. *Actes de la Recherche en Sciences Sociales 74*, 2–22.

Shinn, T. (2000). Formes de division du travail scientifique et convergence intellectuelle: La recherche technico-instrumentale. *Revue française de sociologie 41*(3), 447–473.

Shrum, W. (2005). Reagency of the Internet, or, how I became a guest for science. *Social Studies of Science 35*(5), 723–754. doi:10.1177/0306312705052106

Sierra Iglesias, J. P. (1990). *Salvador Mazza, su vida, su obra: Redescubridor de la enfermedad de Chagas*. Jujuy: Universidad Nacional de Jujuy.

Sigal, S. (1991). *Intelectuales y poder en la década del sesenta*. Buenos Aires: Puntosur Editores.

Sironi, M., Rowntree, V., Snowdon, C., Valenzuela, L., & Marón, C. (2005). Kelp gulls (*Larus dominicanus*) feeding on southern right whales (*Eubalaena australis*) at Península Valdés, Argentina: Updated estimates and conservation implications. *Journal of Cetacean research and Management 61/BRG19*.

Sismondo, S. (2004). *An introduction to science and technology studies*. Malden, MA: Wiley-Blackwell.

Snow, C. P. (1959). *The two cultures*. Cambridge, MA: Cambridge University Press,

Sordelli, A. (1942). *Creación y funcionamiento de una sección para el estudio de virus filtrables*. Buenos Aires: National Institute of Bacteriology Archives.

Sosa Estani, S., & Segura, E. (1999). Tratamiento de la infección por "Trypano-soma Cruzi" en fase indeterminada. Experiencia y normatización actual en la Argentina. *Medicina 59*(Suppl. II), 166–170.

Stengers, I. (1997). *Sciences et pouvoirs*. Paris: La Découverte.

Stent, G. (1968). That was the molecular biology that was. *Science*, New Series *160*(3826), 390–395.

Stepan, N. (1976). *Beginnings of Brazilian science: Oswaldo Cruz, medical research and policy 1890–1920*. New York: Science History Publications and Neale Watson Academic Publications.

Stevens, H. (2013). *Life out of squence: A data-driven history of bioinformatics*. Chicago: University of Chicago Press.

Stichweh, R. (1992). The sociology of scientific disciplines: On the genesis and stability of the disciplinary structure in modern science. *Science in Context 5*(1), 3–15.

Stichweh, R. (1996). Science in the system of world society. *Social Science Information 35*, 327–340.

Storino, R. (2000). La cara oculta de la enfermedad de Chagas. *Revista de la Federación Argentina de Cardiología 29*.

Storino, R. (2002). El tratamiento antiparasitario en la enfermedad de chagas, ¿debe darse a todos o no? En contra. *Revista Argentina de Cardiología 70*(5).

Storino, R., & Milei, J. (1994). *Enfermedad de Chagas*. Buenos Aires: Mosby.

Taborga, A. (2010). *Producción de conocimiento en la periferia de la periferia. Grupos de investigación en física pertenecientes a una universidad del interior argentina. 1990–2005* (PhD Dissertation). FLACSO Argentina, Buenos Aires.

Tarleton, R., & Zhang, L. (1999). Chagas disease etiology: Autoinmunity or parasite persistence? *Parasitology Today 15*(3), 94–99.

Thuillier, P. (1975). *Comment est née la biologie moléculaire: La recherche en biologie moléculaire*. Paris: Editions du Seuil.

Timmermans, S., & Berg, M. (2003). *The gold standard: The challenge of evidence-based medicine and standardization in health care*. Philadelphia, PA: Temple University Press.

Twyman, R. (2002). *What are "model organisms?"* Retrieved June 2018, from http://genome.wellcome.ac.uk/doc_WTD020803.html

Varsavsky, O. (1969). *Ciencia, política, cientificismo*. Buenos Aires: Capital Intelectual.

Velho, L. (2002). North-South collaboration and systems of innovation. *The International Journal of Technology Management and Sustainable Development 1*(3), 171–185.

Velho, L. (2011). La ciencia y los paradigmas de la política científica, tec-nológica y de innovación. In A. Arellano Hernández & P. Kreimer (Eds.), *Estudio social de la ciencia y la tecnología desde América Latina*. Bogotá: Siglo del hombre.

Vermeulen, N. (2013). From Darwin to the census of marine life: Marine biology as big science. *PLoS One 8*(1), e54284. doi:10.1371/journal.pone.0054284

Vessuri, H. (1983). Consideraciones acerca del estudio social de la ciencia. In E. Díaz, Y. Texera, & H. Vessuri (Eds.), *La ciencia periférica*. Caracas: Monte Avila.

Vessuri, H. (1984). The search for a scientific community in Venezuela: From isolation to applied research. *Minerva 22*, 196–235.

Vessuri, H. (1994a). La ciencia académica en América Latina en el siglo XX. *Redes, Revista de Estudios Sociales de la Ciencia 1*(2), 61–62.

Vessuri, H. (1994b). The institutionalization of western science in developing Countries. In J. J. Salomon, F. Sagasti, & C. Sachs (Eds.), *The uncertain quest: Science, technology and development* (pp. 168–200). Tokyo: United Nations University Press.

Vessuri, H. (1996a). Bitter harvest: The growth of a scientific community in Argentina. In J. Gaillard, V. V. Krishna, & R. Waast (Eds.), *Scientific communities in the developing world*. Part 3: Scientific Communities in Latin America (pp. 307–335). New Delhi, Thousand Oaks, CA and London: Sage Publications.

Vessuri, H. (1996b). Scientific cooperation among unequal partners: The strait jacket of the human resource base. In J. Gaillard (Ed.), *Coopérations Scientifiques Internationales* (pp. 171–185). Paris: Éditions de l'ORSTOM.

Vessuri, H. (1996c). Becoming a scientist in Mexico: The challenge of creating a scientific community in an underdeveloped country. *Social Studies of Science 26*(1), 186–191.

Vessuri, H. (2014). "El hombre del maíz de la Argentina": Salomón Horovitz y la tecnología de la investigación en la fitotecnia sudamericana. *Estudios Interdisciplinarios de América Latina y el Caribe 14*(1).

Vessuri, H., & Kreimer, P. (2018). La science latino-américaine: tensions du passé et enjeux du présent. In M. Kleiche-Dray (Ed.), *Les ancrages nationaux de la science mondiale*. Paris: Edition des Archives contemporaines.

Viotti, R., Vigliano, C., Armenti, H., & Segura, E. (1994). Treatment of chronic Chagas' disease with benznidazol: Clinical and serologic evolution of patients with long-term follow-up. *American Heart 127*(1), 151–162.

Viotti, R., Vigliano, C., Lococo, B., Bertocchi, G., Petti, M., Alvarez, G., . . . Armenti, A. (2006). Long-termn cardiac outcomes of treating chronic Chagas disease with Benznidazol versus no treatment: A nonrandomized trial. *Annals of Internal Medicine 144*(10), 724–734.

Vos, R., Horstman, K., & Penders, B. (2008). Walking the line between lab and computation: The "moist" zone. *BioScience 58*(8), 747–755. doi:10.1641/B580811

Wagner, C. S. (2005). Six case studies of international collaboration in science. *Scientometrics 62*(1).

Wagner, C. S. (2006). International collaboration in science and technology: Promises and pitfalls. In L. Box & R. Engelhard (Eds.), *Science and technology policy for development: Dialogues at the interface* (pp. 165–176). London: Anthem Press.

Wagner, C. S. (2008). *The new invisible college: Science for development*. Washington, DC: Brookings Institution Press.

Wagner, C. S., Brahmakulam, I. T., Jackson, B. A., Wong, A., & Yoda, T. (2001). *Science & technology collaboration: Building capacity in developing countries?* Santa Monica, CA: RAND Corporation.

Wagner, C. S., & Kit Wong, S. (2012). Unseen science? Representation of BRICs in global science. *Scientometrics 90*(3), 1001–1013.

Wagner, C. S., Whetsell, T. A., & Leydesdorff, L. (2017). Growth of international collaboration in science: Revisiting six specialties. *Scientometrics 110*(3), 1633–1652.

Wallerstein, I. (1974). *The modern world-system, vol. I: Capitalist agriculture and the origins of the European world-economy in the sixteenth century.* New York, NY and London: Academic Press.

Wetterstrand, K. A. (2015). *DNA sequencing costs: Data from the NHGRI genome sequencing program (GSP).* Retrieved October 2016, from www.genome.gov/sequencingcosts

White, R. W., Gillon, K. W., Black, A. D., & Reid, J. B. (2002). *The distribution of seabirds and marine mammals in Falkland Island waters.* Peterborough: JNCC (Joint Nature Conservation Committee).

Whitley, R. (1972). *Blackboxism* and the sociology of science: A discussion of the major developments in the field. *The Sociological Review Monograph,* (18), 61–92.

Whitley, R. (1976). Umbrella and polytheistic scientific: Disciplines and their elites. *Social Studies of Science 6,* 471–497.

Whitley, R. (1985). *The intellectual and social organization of the sciences.* Oxford and New York, NY: The Clarendon Press and Oxford University Press.

Whitley, R. (2010). Reconfiguring the public sciences: The impact of governance changes on authority and innovation in public science systems. In R. Whitley, J. Glaeser, & L. Engwall (Eds.), *Reconfiguring knowledge production: Changing authority relationships in the sciences and their consequences for intellectual innovation* (pp. 3–48). Oxford: Oxford University Press.

WHO. (1991). *Control de la enfermedad de Chagas: Informe de un Comité de Expertos de la OMS.* Buenos Aires: World Health Organization.

WHO. (2000). *The world health report.* Geneva: World Health Organization.

WHO/TDR. (2005). *Vision et stratégie pour les dix ans à venir.* Geneva: Programme Spécial de recherche et de développement concernant les maladies Tropicales (TDR).

Winner, L. (1993). Upon opening the black box and finding it empty: Social constructivism and the philosophy of technology. *Science, Technology and Human Values 18*(3), 362–378.

Zabala, J. P. (2010). *La enfermedad de Chagas en la Argentina. Investigación científica, problemas sociales y políticas sanitarias.* Buenos Aires: Editorial de la Universidad Nacional de Quilmes.

Index

For Product Safety Concerns and Information please contact our EU
representative GPSR@taylorandfrancis.com
Taylor & Francis Verlag GmbH, Kaufingerstraße 24, 80331 München, Germany

www.ingramcontent.com/pod-product-compliance
Lightning Source LLC
Chambersburg PA
CBHW060347220326
41598CB00023B/2828